# Quantum Fermentation

## Microbial Dynamics in Wine Production

**Zera Norman**

# Chapter 1: The Microbial Frontier in Winemaking

## Microbes as Catalysts of Flavor

The unsung heroes of the winemaking process lie in the invisible realm of microbiology. While casual wine enthusiasts may marvel at the complexities of vintage and terroir, the true alchemists responsible for transforming grape juice into the nectar of the gods are the diverse array of microscopic organisms that inhabit every step of the fermentation journey. From the vineyard to the barrel, these microscopic catalysts shape the flavors, aromas, and textures that elevate wine to the status of art.

At the heart of this microbial alchemy lies the concept of terroir - the unique blend of environmental factors, from soil composition to climate, that imbue a wine with its distinct regional identity. What many fail to appreciate, however, is that terroir is not merely a passive influence, but an intricate dance between the vines and their invisible microbial partners. The native yeasts and bacteria that thrive in a particular viticultural region act as living conduits, translating the nuances of the local ecosystem into the finished product.

Consider the case of Brettanomyces, a genus of yeast with a reputation that evokes both reverence and revulsion among wine connoisseurs. While certain strains can impart unsavory "barnyard" aromas if left unchecked, the judicious use of Brettanomyces can lend a wine an alluring complexity, contributing notes of leather, smoke, and even the elusive "animal" character that oenophiles prize. This microbial maverick is but one example of how the unseen world of microorganisms can shape a wine's sensory profile, adding depth and individuality to even the most celebrated appellations.

Beyond the contributions of specific microbes, the overall diversity of the vineyard microbiome plays a critical role in defining a wine's signature. Just as a symphony orchestra requires a harmonious blend of instruments to achieve its full emotional impact, the microbial community of a given terroir acts as a living orchestra, each member contributing a unique voice to the symphony of flavors. When this delicate balance is disrupted - whether through the use of synthetic pesticides, monoculture farming practices, or other ecological disruptions - the results can be catastrophic, leading to a homogenization of flavors and a loss of the very characteristics that wine lovers cherish.

Fortunately, a growing number of winemakers are recognizing the pivotal role of microbes in crafting wines of remarkable personality and finesse. By embracing the inherent biodiversity of their vineyards, these visionary producers are learning to harness the creative potential of the microbial world, using techniques like wild fermentation, extended maceration, and judicious use of indigenous yeasts to coax out the most captivating flavors and aromas. The results are wines that speak eloquently of their origins, each sip a unique expression of the complex interplay between vine, soil, and the invisible legions of microbial artisans.

As our understanding of the wine microbiome continues to evolve, the future of winemaking promises to be an ever more collaborative endeavor, one in which humans and microbes work in harmonious synergy to elevate the art of fermentation to new heights. In this brave new world of microbial oenology, the boundaries between the visible and invisible realms will continue to blur, revealing the profound truth that the most transcendent wines are not the products of human ingenuity alone, but the collective efforts of all the living organisms that come together to create them.

## Terroir and the Microbiome

The concept of terroir has long captivated the imaginations of wine enthusiasts, conjuring visions of sun-drenched vineyards and ancient soils imbued with the spirit of place. Yet for all the romantic notions that surround this French term, the true essence of terroir resides not in the physical landscape alone, but in the intricate web of microbial life that thrives within it.

To understand the role of the microbiome in expressing a wine's terroir, one must first grasp the profound interconnectedness that exists between the visible and invisible realms of the vineyard ecosystem. Like a symphony orchestra, where each instrument contributes its unique voice to the greater whole, the myriad microorganisms that inhabit the soil, the grape skins, and the fermentation vessels work in concert to translate the nuances of a particular growing region into the flavors, aromas, and textures that distinguish one wine from another.

Consider, for example, the case of Burgundy, a hallowed winemaking region renowned for the transcendent elegance of its Pinot Noir and Chardonnay. The soils of Burgundy, carved from ancient seabeds and imbued with fossilized marine life, harbor a rich and diverse microbiome that helps to shape the distinctive minerality and complexity that define the region's most celebrated cuvées. The yeasts and bacteria that thrive in these calcareous soils do not simply ferment the grape sugars into alcohol; rather, they act as living catalysts, unlocking a symphony of volatile compounds that contribute to a wine's aromatic profile, from the haunting perfume of violets to the savory umami notes that linger on the palate.

This dynamic interplay between terroir and microbiome is not limited to prestigious appellations, but can be observed across the winemaking world. In the Mosel valley of Germany, for instance, the steep, slate-laden slopes foster a microbial community that imbues the region's Rieslings with a riveting tension between fruit and minerality. Conversely, the warm, sun-drenched hills of Tuscany nurture a microbial terroir that lends the Sangiovese grape an alluring rusticity, with notes of dried herbs, spice, and leather.

Understanding the role of the microbiome in expressing terroir is not merely an academic exercise, but a crucial consideration for vintners seeking to craft wines of unparalleled authenticity and character. By embracing the inherent biodiversity of their vineyards and learning to work in harmony with the invisible legions of microbial artisans, these visionary producers are able to coax out the most captivating flavors and aromas, crafting wines that serve as tangible extensions of their unique growing environments.

Yet the relationship between terroir and microbiome is not without its challenges. In an age of industrialized agriculture and globalized commerce, the delicate balance of the vineyard microbiome has come under increasing threat, as monoculture farming practices, synthetic pesticides, and other ecological disruptions have led to a homogenization of microbial diversity. The consequences of this microbial impoverishment can be devastating, as wines lose their sense of place and become mere reflections of standardized techniques rather than true expressions of terroir.

Fortunately, a growing movement of environmentally conscious winemakers is working to restore the integrity of the vineyard microbiome, embracing organic and biodynamic farming methods that nurture the complex web of microbial life. By cultivating a thriving, biodiverse ecosystem in their vineyards, these producers are able to craft wines that capture the essence of their unique growing environments, offering wine lovers a visceral connection to the land and the invisible legions that bring it to life.

As our understanding of the wine microbiome continues to evolve, the future of viticulture promises to be an ever more collaborative endeavor, one in which humans and microbes work in harmonious synergy to elevate the art of winemaking to new heights. In this brave new world of microbial oenology, the boundaries between the visible and invisible realms will continue to blur, revealing the profound truth that the most transcendent wines are not the products of human ingenuity alone, but the collective efforts of all the living organisms that come together to create them.

## Harnessing Microbial Diversity

The key to unlocking the true potential of wine lies not in the physical terroir alone, but in the intricate web of microbial life that pervades every aspect of the viticultural landscape. Just as a symphony orchestra requires a harmonious blend of instruments to achieve its full emotional impact, the microbial community of a given growing region acts as a living orchestra, each member contributing a unique voice to the symphony of flavors and aromas that distinguish the world's most captivating vintages.

Yet in an age of industrialized agriculture and globalized commerce, this delicate microbial balance has come under increasing threat. As monoculture farming practices, synthetic pesticides, and other ecological disruptions have homogenized the vineyard microbiome, many winemakers have found themselves struggling to capture the essence of their terroir, their wines reduced to mere reflections of standardized techniques rather than true expressions of place.

Fortunately, a growing number of visionary producers are recognizing the vital role that microbial diversity plays in crafting wines of remarkable personality and finesse. By embracing the inherent biodiversity of their vineyards and learning to work in harmony with the invisible legions of microbial artisans, these trailblazers are ushering in a new era of oenological excellence, one in which the boundaries between the visible and invisible realms continue to blur.

Consider the case of Lalou Bize-Leroy, the legendary Burgundian vigneron whose meticulous biodynamic practices have earned her wines a near-mythical status among connoisseurs. Through her unwavering commitment to nurturing the vineyard microbiome, Bize-Leroy has coaxed out the most captivating expressions of Pinot Noir and Chardonnay, her Domaine Leflaive and Domaine Leroy bottlings brimming with an ethereal complexity that transcends the constraints of vintage variation.

The key to Bize-Leroy's success lies in her deep understanding of the symbiotic relationship between vine and microbe. By eschewing synthetic inputs in favor of holistic, soil-building techniques, she has fostered a thriving, biodiverse ecosystem in her vineyards, enabling a vast array of yeast and bacterial species to flourish and contribute their unique voices to the finished wines. The resulting cuvées are living tapestries, woven from the collective efforts of countless microscopic artisans, each one playing its part in the creation of something greater than the sum of its parts.

Nor is Bize-Leroy's approach limited to the hallowed terroirs of Burgundy. Across the winemaking world, a new generation of producers is embracing the power of microbial diversity, using techniques like wild fermentation, extended maceration, and judicious use of indigenous yeasts to coax out the most compelling expressions of their unique growing environments. In California's Sonoma Coast, for instance, the Peay family has cultivated a thriving microbial ecosystem in their biodynamically farmed vineyards, yielding Pinot Noirs and Chardonnays that captivate the senses with their haunting minerality and ethereal elegance.

As our understanding of the wine microbiome continues to deepen, the future of viticulture promises to be an ever more collaborative endeavor, one in which humans and microbes work in harmonious synergy to elevate the art of fermentation to new heights. Through the skillful harnessing of microbial diversity, winemakers will unlock increasingly complex and nuanced flavor profiles, crafting wines that serve as visceral conduits to the living terroirs from whence they spring.

Indeed, the true mastery of winemaking lies not in the imposition of human will, but in the humble acknowledgment that we are but stewards of a process far greater than ourselves. By recognizing the vital role that microbes play in shaping the character of a wine, we open ourselves to a world of infinite creative potential, where the unseen becomes the wellspring of the sublime. In this brave new era of microbial oenology, the most captivating vintages will be those that celebrate the symphony of life that courses through every glass, a testament to the profound interconnectedness that binds us all.

## Emerging Trends in Microbial Enology

The once-obscure realm of microbial oenology is undergoing a transformative renaissance, as visionary winemakers and cutting-edge researchers unlock the hidden potential of the vineyard microbiome. Driven by an insatiable curiosity and a steadfast commitment to quality, these trailblazers are ushering in a new era of flavor innovation, one in which the invisible legions of microscopic artisans take center stage in the creation of the world's most captivating vintages.

At the heart of this microbial revolution lies a fundamental shift in our understanding of the winemaking process. Gone are the days when fermentation was viewed as a purely technical exercise, a matter of methodically controlling temperature, yeast selection, and sulfur additions. Today, the most forward-thinking producers recognize that the true magic of wine lies not in the imposition of human will, but in the careful cultivation of the diverse microbial communities that thrive in their vineyards and cellars.

Consider the case of Emilien Boutillat, the dynamic young winemaker behind Clos Roche Blanche in France's Loire Valley. Eschewing the standardized approaches that have come to dominate modern oenology, Boutillat has embraced the power of wild fermentation, allowing the native yeasts and bacteria that cling to his Sauvignon Blanc and Chenin Blanc grapes to guide the transformation from mere grape juice to transcendent nectar. The resulting wines - vibrant, mineral-driven, and imbued with a captivating array of aromatic nuances - stand in stark contrast to the overly homogenized examples that so often grace the shelves of wine shops and restaurants.

Nor is Boutillat's approach limited to the hallowed terroirs of France. Across the globe, a new generation of microbial mavericks is pushing the boundaries of what is possible in the realm of fermented grape juice. In California's Sonoma Coast, for instance, the team at Peay Vineyards has harnessed the power of indigenous yeast strains to craft Pinot Noirs and Chardonnays that deftly balance ripe fruit and bracing acidity, their complex aromatic profiles evocative of the fog-shrouded coastal hills from which they hail.

Meanwhile, in the Okanagan Valley of British Columbia, the vintners at Okanagan Crush Pad have taken their fascination with microbial diversity to bold new heights, crafting multi-vintage, solera-style blends that showcase the kaleidoscopic interplay of yeasts, bacteria, and time. These unique bottlings - part living, breathing organism, part oenological alchemy - defy easy categorization, challenging the very notion of what a "finished" wine should be.

As our understanding of the wine microbiome continues to evolve, the future of microbial enology promises to be one of ever-greater complexity and creativity. Through the application of cutting-edge analytical tools, from next-generation sequencing to real-time metabolomic profiling, winemakers will gain unprecedented insight into the invisible world that underpins their craft, empowering them to fine-tune their practices and unlock ever-more-nuanced expressions of terroir.

Equally transformative will be the emergence of microbial interventions designed to enhance the flavor and quality of wine. Building upon the success of probiotic and synbiotic approaches in other fields, pioneering oenologists are exploring the use of custom-tailored yeast and bacterial consortia to address challenges ranging from stuck fermentations to the mitigation of undesirable aromatic compounds. By harnessing the power of engineered microbial communities, these innovators are poised to usher in a new era of precision winemaking, where the invisible legions of the cellar become not merely tools, but active collaborators in the creative process.

Indeed, as the frontiers of microbial enology continue to expand, the very notion of what constitutes a "wine" may undergo a radical redefinition. Already, we are witnessing the emergence of radical new approaches, from the use of unconventional substrates (think: fruit, grains, even seaweeds) to the incorporation of cutting-edge biotechnologies like quantum computing and artificial intelligence. In this brave new world, the boundaries between the natural and the engineered, the visible and the invisible, will continue to blur, ushering in a renaissance of oenological innovation that will captivate and delight wine lovers for generations to come.

# Chapter 2: The Quantum Nature of Fermentation

## Quantum Mechanics and the Microbial World

The boundaries between the visible and invisible realms have long captivated the human imagination. In the grand scheme of the cosmos, we find ourselves perched precariously between the unfathomable vastness of the universe and the inscrutable realm of the subatomic - a liminal space where the laws of classical physics give way to the strange and often counterintuitive principles of quantum mechanics. Yet it is within this quantum frontier that we find the key to unlocking the mysteries of the microbial world, the unseen legions that shape the very fabric of our winemaking universe.

Consider, for a moment, the curious case of Saccharomyces cerevisiae, the stalwart yeast species that has served as the workhorse of the fermentation industry for centuries. On the surface, this humble microbe might appear to be a straightforward, predictable entity, dutifully converting grape sugars into alcohol and carbon dioxide. But delve deeper into the quantum nature of its cellular machinery, and a far more complex and enchanting reality emerges.

At the subatomic level, the metabolic processes that power the yeast cell operate according to the strange and often counterintuitive rules of quantum mechanics. Electrons, the fundamental building blocks of matter, do not simply follow the deterministic pathways prescribed by classical physics, but rather exist in a state of quantum superposition - inhabiting multiple possible states simultaneously until the moment of observation. It is this quantum uncertainty, this inherent unpredictability, that imbues the yeast cell with a dynamism and adaptability that confounds our linear, macroscopic understanding of the natural world.

Nowhere is this quantum influence more apparent than in the realm of microbial communication and cooperation. Cutting-edge research has revealed that the microbes populating a wine's fermentation vessel do not simply act as isolated individuals, but rather engage in a complex, entangled network of information exchange. Through the transmission of quantum-mediated signals, these microscopic entities coordinate their metabolic activities, synchronizing their rhythms in a symphony of fermentative alchemy.

This quantum entanglement extends beyond the boundaries of the individual cell, binding the myriad microorganisms that make up the living tapestry of the wine microbiome. Just as the instruments of an orchestra must harmonize to produce a transcendent musical experience, so too must the diverse array of yeasts, bacteria, and fungi that inhabit a wine's terroir work in concert, their quantum-level interactions shaping the flavors, aromas, and textures that distinguish the world's most captivating vintages.

It is a humbling realization, this notion that the very essence of a wine's identity is rooted in the quantum realm - a domain that defies our most intuitive conceptions of cause and effect, of determinism and control. And yet, for the visionary winemakers who have come to embrace the quantum nature of fermentation, this newfound understanding has unlocked a world of creative potential, empowering them to cultivate ever-more-nuanced and compelling expressions of terroir.

Through the judicious application of quantum-inspired techniques, such as the strategic manipulation of environmental conditions to induce desired microbial behaviors, these trailblazers are able to coax out the most captivating flavors and aromas, crafting wines that serve as tangible extensions of their unique growing environments. By harnessing the inherent unpredictability and interconnectedness of the microbial world, they transcend the limitations of standardized oenological practices, ushering in a new era of microbial alchemy that defies easy categorization.

As our comprehension of the quantum underpinnings of the microbial world continues to evolve, the future of winemaking promises to be one of ever-greater complexity and creativity. The boundaries between the visible and invisible realms will continue to blur, revealing the profound truth that the most transcendent wines are not the products of human ingenuity alone, but the collective efforts of all the living organisms that come together to create them. In this brave new world of quantum enology, the possibilities are as endless as the stars themselves, waiting to be uncovered by those bold enough to peer into the shimmering, uncertainty-laden heart of the microbial frontier.

# Entanglement and the Interconnectedness of Microbes

At the very heart of the microbial universe lies a profound and often counterintuitive truth: the uncountable legions of microscopic entities that inhabit our world are not merely isolated individuals, but rather a vast, interconnected web of living information. Like the celestial bodies that dot the night sky, these microscopic denizens of the wine cellar and vineyard exist in a state of quantum entanglement, their fates intricately entwined in ways that defy our most cherished notions of cause and effect.

Consider, for a moment, the curious case of Oenococcus oeni, the bacterial species responsible for the crucial malolactic fermentation stage in many celebrated wine styles. On the surface, this unassuming microbe might appear to be a straightforward, predictable player in the oenological drama, dutifully converting the tart malic acid into softer lactic acid, thereby smoothing the wine's mouthfeel and enhancing its complexity. But delve deeper into the quantum nature of its cellular machinery, and a far more enchanting reality emerges.

Through the transmission of quantum-mediated signals, Oenococcus oeni engages in a complex choreography of communication and cooperation with the other microbial inhabitants of the winemaking environment. Its metabolic activities are not merely the product of isolated, deterministic processes, but rather the result of an intricate, entangled network of information exchange - a symphony of fermentative alchemy in which each player's actions reverberate through the entire microbial community, shaping the ultimate character of the wine.

And Oenococcus oeni is but one example of this quantum interconnectedness. Throughout the viticultural landscape, from the vine-cloaked slopes of Bordeaux to the fog-shrouded hills of California's Sonoma Coast, the microbes that call these terroirs home exist in a state of quantum entanglement, their collective behaviors and metabolic outputs woven into the very fabric of the wines they help to create.

It is a humbling realization, this notion that the essence of a wine's identity is not the sole province of human intervention, but rather the result of a delicate dance between the visible and invisible realms. For the visionary winemakers who have come to embrace the quantum nature of fermentation, however, this newfound understanding has unlocked a world of creative potential, empowering them to cultivate ever-more-nuanced and compelling expressions of place.

By nurturing the inherent biodiversity of their vineyards and cellars, these trailblazers create the conditions for a thriving, entangled microbial community to emerge - one in which the diverse array of yeasts, bacteria, and fungi work in harmonious synergy, their quantum-level interactions shaping the flavors, aromas, and textures that distinguish the world's most captivating vintages. Through the judicious application of techniques like wild fermentation, extended maceration, and the selective reintroduction of indigenous microbes, they coax out the most captivating expressions of terroir, each sip a tangible manifestation of the quantum orchestra that sings beneath the surface.

Yet the path to microbial harmony is not without its challenges. In an age of industrialized agriculture and globalized commerce, the delicate balance of the vineyard microbiome has come under increasing threat, as monoculture farming practices, synthetic pesticides, and other ecological disruptions have led to a homogenization of microbial diversity. The consequences of this microbial impoverishment can be devastating, as wines lose their sense of place and become mere reflections of standardized techniques rather than true expressions of the quantum entanglement that lies at the heart of oenological excellence.

Fortunately, a growing movement of environmentally conscious winemakers is working to restore the integrity of the vineyard microbiome, embracing organic and biodynamic farming methods that nurture the complex web of quantum-entangled life. By cultivating a thriving, biodiverse ecosystem in their vineyards and cellars, these visionaries are able to craft wines that capture the essence of their unique growing environments, offering wine lovers a visceral connection to the land and the invisible legions that bring it to life.

As our understanding of the quantum nature of the microbial world continues to evolve, the future of winemaking promises to be an ever more collaborative endeavor, one in which humans and microbes work in harmonious synergy to elevate the art of fermentation to new heights. In this brave new era of quantum enology, the boundaries between the visible and invisible realms will continue to blur, revealing the profound truth that the most transcendent wines are not the products of human ingenuity alone, but the collective efforts of all the living organisms that come together to create them.

## Quantum Tunneling and Microbial Metabolism

The intricate dance of microbial life that courses through the veins of the winemaking world is a symphony of unparalleled complexity, driven by the strange and often counterintuitive principles that govern the quantum realm. At the heart of this microbial alchemy lies the phenomenon of quantum tunneling, a perplexing process by which subatomic particles break free of the constraints of classical physics, transcending the barriers that would otherwise impede their movement.

Consider, for a moment, the case of Brettanomyces, a genus of yeast that has long vexed and captivated wine enthusiasts in equal measure. To the casual observer, the metabolic activities of this microscopic maverick may appear straightforward enough - the conversion of grape sugars into alcohol and a host of volatile aromatic compounds. Yet delve deeper into the quantum-level mechanics that drive Brettanomyces, and a far more enchanting reality emerges.

Through the process of quantum tunneling, the enzymes that power Brettanomyces' metabolic pathways are able to overcome energetic barriers that would traditionally be insurmountable, catalyzing reactions and transformations that defy our most intuitive conceptions of cause and effect. By tunneling through the potential energy landscapes that define the biochemical landscape, these quantum-enabled enzymes are able to access previously inaccessible states, unlocking a veritable cornucopia of flavor compounds that impart Brettanomyces-influenced wines with their distinctive "barnyard" or "leather" aromas.

Nor is Brettanomyces the sole beneficiary of quantum tunneling's creative potential. Throughout the microbial world that inhabits the winemaking environment, from the stalwart Saccharomyces cerevisiae to the enigmatic Oenococcus oeni, these quantum-enabled metabolic processes play a pivotal role in shaping the flavors, aromas, and textures that distinguish the world's most captivating vintages.

Consider, for instance, the case of Oenococcus oeni and its quantum-assisted conversion of malic acid to lactic acid during the crucial malolactic fermentation stage. By tunneling through the energetic barriers that typically impede this transformation, the bacterial enzymes responsible are able to catalyze the reaction with far greater efficiency, unlocking a cascade of downstream effects that enhance a wine's mouthfeel, complexity, and overall harmony.

It is a humbling realization, this notion that the very essence of a wine's identity is rooted in the quantum realm - a domain that defies our most cherished notions of determinism and control. And yet, for the visionary winemakers who have come to embrace the quantum nature of fermentation, this newfound understanding has unlocked a world of creative potential, empowering them to cultivate ever-more-nuanced and compelling expressions of terroir.

Through the judicious application of quantum-inspired techniques, such as the strategic manipulation of environmental conditions to induce desired microbial behaviors, these trailblazers are able to coax out the most captivating flavors and aromas, crafting wines that serve as tangible extensions of their unique growing environments. By harnessing the inherent unpredictability and interconnectedness of the microbial world, they transcend the limitations of standardized oenological practices, ushering in a new era of microbial alchemy that defies easy categorization.

As our comprehension of the quantum underpinnings of the microbial world continues to evolve, the future of winemaking promises to be one of ever-greater complexity and creativity. The boundaries between the visible and invisible realms will continue to blur, revealing the profound truth that the most transcendent wines are not the products of human ingenuity alone, but the collective efforts of all the living organisms that come together to create them. In this brave new world of quantum enology, the possibilities are as endless as the stars themselves, waiting to be uncovered by those bold enough to peer into the shimmering, uncertainty-laden heart of the microbial frontier.

# Superposition and the Uncertainty of Fermentation

The dance between certainty and uncertainty in fermentation processes mirrors the quantum mechanical principle of superposition, where multiple states exist simultaneously until observation collapses them into a single outcome. Traditional fermentation practices, passed down through generations, have always acknowledged this inherent unpredictability while working to harness it productively.

When we initiate a fermentation process, countless microorganisms begin their metabolic activities, creating a complex ecosystem where multiple reactions occur simultaneously. Much like Schrödinger's famous thought experiment, we cannot precisely predict which strains will dominate or how their interactions will unfold until we observe the results. This uncertainty is not a weakness but rather a fundamental characteristic that gives fermented foods their unique and complex flavors.

Consider a traditional sourdough starter: at any given moment, various strains of wild yeast and lactic acid bacteria exist in different proportional states. The final flavor profile emerges from this microbial superposition, influenced by temperature, humidity, flour composition, and countless other environmental factors. Master bakers understand this principle intuitively, adjusting their techniques based on subtle cues rather than rigid formulas.

The uncertainty principle manifests clearly in natural wine making, where indigenous yeasts present on grape skins compete for resources. Unlike controlled industrial fermentation, these natural processes embrace variability. Each vintage represents a unique collapse of possibilities into reality, influenced by factors ranging from weather patterns to soil microbiota. This explains why natural wines can display such remarkable vintage variation, each bottle telling its own story of microbial interaction and environmental influence.

Temperature plays a crucial role in this superposition of states. A mere two-degree difference can shift the balance between different bacterial strains in kimchi fermentation, leading to distinctly different flavor outcomes. Korean masters of kimchi-making account for this by adjusting salt levels and fermentation times based on subtle environmental cues, working with rather than against this uncertainty.

The same principles apply to cheese aging, where multiple strains of bacteria and fungi coexist in various states of activity. The complex flavors of aged cheeses emerge from this sustained period of microbial superposition, where different organisms dominate at different times. Cave-aged cheeses particularly exemplify this concept, as the natural environment maintains conditions where multiple microbial states can coexist productively.

Modern fermentation practitioners can learn from this understanding of superposition. Instead of trying to eliminate all variables, we can work within the bounds of uncertainty while maintaining food safety. This might mean embracing batch-to-batch variation in kombucha brewing or accepting that each crock of sauerkraut will develop its own unique character.

pH monitoring offers a window into this dynamic system, much like quantum measurements collapse wave functions into definite states. Regular pH readings can help track fermentation progress while still allowing for natural variation within safe parameters. This balance between control and uncertainty defines successful fermentation practice.

The time dimension adds another layer of complexity to fermental superposition. Short-term fermentations like yogurt making might seem more predictable, but even here, multiple bacterial strains work in concert, their relative activities shifting throughout the process. Longer fermentations, such as traditional miso or aged vinegar, demonstrate how these microbial ecosystems evolve over months or years.

Understanding fermental superposition helps explain why scaling up traditional fermentation processes often proves challenging. Industrial production seeks consistency through strict control, potentially sacrificing the complex flavors that emerge from natural variation. This tension between standardization and traditional methods continues to shape modern fermentation practices.

Temperature gradients within fermentation vessels create micro-environments where different organisms can thrive, contributing to the overall complexity of the final product. This spatial variation, combined with temporal changes, creates a four-dimensional matrix of possibility that collapses into our experienced reality of taste and texture.

The practical application of this understanding leads to more intuitive fermentation practices. Rather than fighting against uncertainty, we can embrace it while maintaining appropriate safety boundaries. This might mean allowing temperature fluctuations within a specific range or accepting that each batch of fermented vegetables will develop its own unique character.

The relationship between fermentation uncertainty and final product quality remains complex. Sometimes, the most interesting and delicious results emerge from processes that seemed initially uncertain or even concerning. This parallels the quantum understanding that uncertainty is not merely a limitation of measurement but a fundamental property of reality itself.

## Quantum Coherence and the Orchestration of Microbial Processes

Microbial communities operate with a remarkable degree of coordination that bears striking similarities to quantum coherent systems. The synchronized behaviors of bacteria, yeasts, and other microorganisms during fermentation processes demonstrate patterns that suggest deeper underlying principles of collective organization.

When billions of microorganisms work in concert during fermentation, they exhibit behaviors analogous to quantum coherence, where particles maintain synchronized states across relatively large distances. This coordination becomes evident in the way different strains of bacteria communicate through chemical signaling, creating complex networks of interaction that influence the entire fermentation process.

The metabolic pathways in fermentation display coherent patterns that optimize energy flow and resource utilization. Much like quantum systems that find the most efficient routes through complex energy landscapes, microbial communities self-organize to maximize their collective survival and productivity. This optimization occurs without central control, emerging from countless local interactions between individual organisms.

Temperature fluctuations play a crucial role in maintaining this coherent behavior. Just as quantum systems require specific conditions to maintain coherence, microbial communities need precise environmental parameters to sustain their coordinated activities. Small variations in temperature can ripple through the entire system, affecting the synchronized behaviors of countless organisms.

The phenomenon becomes particularly visible in pellicle formation during kombucha fermentation, where cellulose-producing bacteria create intricate networks that span the entire surface of the liquid. This self-organizing structure demonstrates how individual microbes can contribute to larger, coherent patterns that serve the entire community.

Chemical gradients within fermentation vessels create zones of varying activity, yet these zones maintain remarkable coordination. The way different species position themselves along these gradients shows evidence of collective intelligence, similar to how quantum particles distribute themselves across energy levels in coherent systems.

Timing in fermentation processes often displays surprising precision, with different stages of metabolic activity occurring in coordinated sequences. This temporal coherence suggests underlying mechanisms that help maintain synchronization across large populations of microorganisms, similar to quantum coherence in molecular systems.

The exchange of genetic material between microbes during fermentation adds another layer of coherent behavior. Horizontal gene transfer allows beneficial traits to spread rapidly through the population, creating adaptive responses that benefit the entire community. This sharing of information parallels the way quantum systems can transmit information across entangled particles.

pH changes during fermentation demonstrate how local actions can have system-wide effects. When acid-producing bacteria lower the pH, they create conditions that influence the entire microbial ecosystem. This coordinated response to environmental changes shows how individual actions contribute to collective adaptation.

Biofilm formation in fermented foods represents another example of quantum-like coherent behavior. These complex bacterial communities develop structured architectures that optimize nutrient flow and protect against environmental stresses. The organization of these biofilms suggests sophisticated communication and coordination mechanisms among participating microbes.

The role of water in maintaining microbial coherence cannot be overstated. Water molecules provide the medium through which chemical signals travel, enabling the synchronized behaviors observed in fermentation. The structured water layers around microorganisms may even facilitate quantum-like information transfer between cells.

Stress responses in microbial communities show remarkable coordination. When environmental conditions become challenging, different species often respond in complementary ways that help maintain the stability of the entire system. This collective response to stress demonstrates how coherent behavior can enhance community resilience.

The production of secondary metabolites during fermentation often follows patterns that suggest underlying coordination. Different strains of microorganisms produce compounds that benefit not just themselves but the entire community, showing how coherent behavior can lead to mutual advantage.

Understanding these coherent processes has practical implications for fermentation control. By recognizing the importance of maintaining conditions that support coordinated microbial activity, we can better guide fermentation outcomes while respecting the natural tendencies of these complex systems.

The spatial organization of microbial communities in fermented foods often displays fractal patterns, suggesting deeper principles of self-organization.

These patterns emerge from the coherent interactions of countless individuals, creating structures that enhance the resilience and functionality of the entire system.

Energy flow through fermentation systems shows evidence of optimization that parallels quantum processes. The way microorganisms distribute resources and coordinate metabolic activities suggests principles of collective organization that transcend simple chemical reactions, pointing toward more fundamental patterns of coherent behavior in biological systems.

# Chapter 3: Mapping the Microbial Landscape

## Advances in Metagenomic Analysis

Metagenomic analysis has revolutionized our understanding of microbial communities, offering unprecedented insights into the complex world of fermentation. The ability to sequence and analyze entire microbial populations simultaneously has transformed traditional fermentation practices into data-driven processes, revealing previously unknown interactions and dependencies.

Recent technological breakthroughs in high-throughput sequencing have enabled researchers to capture the complete genetic diversity present in fermented foods. What once appeared as a simple transformation of ingredients now reveals itself as an intricate dance of thousands of microbial species, each contributing its unique genetic signature to the final product.

The application of metagenomic techniques to traditional fermented foods has unveiled surprising discoveries. For instance, studies of traditional kimchi have identified over 100 bacterial species participating in the fermentation process, with different strains dominating at various stages. This dynamic succession of microbial populations explains the complex flavor development that occurs over time.

Modern sequencing platforms can now process samples in a matter of hours, generating millions of

DNA sequences that paint a detailed picture of microbial ecosystems. This rapid analysis allows for real-time monitoring of fermentation processes, enabling precise interventions when needed to maintain optimal conditions for desired outcomes.

The integration of bioinformatics tools has made it possible to reconstruct complete metabolic pathways from metagenomic data. These reconstructions reveal how different microorganisms collaborate to break down complex compounds and produce the distinctive flavors, aromas, and textures characteristic of fermented foods.

Environmental factors significantly influence microbial community composition, and metagenomic analysis helps track these changes. Temperature fluctuations, pH levels, and substrate availability all leave their mark on the genetic profile of the fermenting population, providing valuable insights for process optimization.

Comparative metagenomic studies of traditional versus industrial fermentations have highlighted the importance of microbial diversity. Traditional methods often harbor richer microbial communities, contributing to more complex flavor profiles and potentially enhanced nutritional benefits.

The discovery of novel enzymes through metagenomic screening has opened new possibilities for fermentation control. These enzymes, evolved over countless generations in traditional fermented foods, offer potential applications in improving industrial fermentation processes and developing new food products.

Temporal analysis of metagenomic data reveals the succession patterns of different microbial species throughout the fermentation process. This understanding helps predict and control the development of desired characteristics while preventing the growth of unwanted organisms.

Machine learning algorithms applied to metagenomic data sets have identified patterns that correlate specific microbial combinations with desirable product attributes. This knowledge enables more precise control over fermentation outcomes while maintaining the complexity that makes traditional fermented foods unique.

The study of horizontal gene transfer within fermentation communities has revealed how beneficial traits spread among different species. This genetic exchange contributes to the adaptation and resilience of fermentation ecosystems, ensuring their stability across varying conditions.

Metagenomic analysis has also illuminated the role of bacteriophages in fermentation processes. These viral particles influence bacterial population dynamics and can significantly impact fermentation outcomes, adding another layer of complexity to our understanding of these systems.

The investigation of rare and understudied microorganisms through metagenomics has expanded our appreciation for biodiversity in fermented foods. Previously overlooked species often play crucial roles in maintaining ecosystem stability and contributing to product characteristics.

Advanced sampling techniques coupled with metagenomic analysis have revealed spatial patterns in microbial distribution within fermentation vessels. This spatial organization influences nutrient flow and metabolic interactions, affecting the final product quality.

The integration of metabolomic data with metagenomic analysis provides a more complete picture of fermentation processes. This combined approach helps identify key metabolic pathways and their regulation, leading to better process control and product consistency.

Recent developments in long-read sequencing technologies have improved our ability to reconstruct complete microbial genomes from complex communities. This advancement enables better understanding of strain-level variations and their impact on fermentation outcomes.

The application of metagenomics to starter culture development has led to more robust and effective fermentation initiators. By understanding the genetic basis of desired traits, researchers can select and combine strains for optimal performance while maintaining traditional characteristics.

These advances in metagenomic analysis continue to bridge the gap between traditional fermentation wisdom and modern scientific understanding. As sequencing technologies become more accessible and analytical tools more sophisticated, our ability to harness and enhance the power of fermentation grows exponentially.

# Profiling the Wine Microbiome

Wine fermentation harbors one of the most complex and dynamic microbiomes in the food world, with countless species of yeasts and bacteria working in concert to transform grape juice into a sophisticated beverage. The intricate dance of microorganisms begins on the grape skin itself, where indigenous species establish the foundation for fermentation long before harvest.

Recent studies have revealed that a single grape berry can host up to 50 different species of yeasts and hundreds of bacterial strains. These microorganisms vary significantly based on geographical location, climate conditions, and vineyard management practices. The terroir concept, long celebrated by wine enthusiasts, extends beyond soil composition to encompass this unique microbial fingerprint of each vineyard.

Saccharomyces cerevisiae, while often considered the primary wine yeast, represents only a small fraction of the initial grape microbiome. Non-Saccharomyces yeasts such as Hanseniaspora, Candida, and Metschnikowia species dominate the early stages of fermentation, contributing distinctive aromatic compounds and flavor precursors that influence the final wine character.

Temperature fluctuations during fermentation create selective pressures that drive microbial succession patterns. As alcohol levels rise, less tolerant species decline, allowing S. cerevisiae to dominate the later stages. This natural selection process creates a temporal progression of flavors and aromas that contribute to wine complexity.

Malolactic fermentation, conducted primarily by Oenococcus oeni, represents another crucial phase in wine microbiome development. These specialized bacteria transform harsh malic acid into softer lactic acid, while simultaneously producing compounds that enhance mouthfeel and contribute buttery notes to certain wines.

The vineyard environment significantly influences wine microbiome composition. Organic and biodynamic practices often result in more diverse microbial communities compared to conventional farming methods. This increased biodiversity can lead to more complex flavor profiles and greater vintage variation.

Modern DNA sequencing techniques have revealed previously unknown participants in wine fermentation. Various species of Metschnikowia, for instance, produce enzymes that release aromatic compounds from grape precursors, contributing significantly to varietal character expression.

Bacterial communities in wine extend beyond the well-known malolactic fermenters. Acetic acid bacteria, while often considered spoilage organisms, can contribute positively to wine complexity when present in controlled numbers. Their activity helps shape the wine's acid profile and influences aging potential.

The interaction between different yeast species creates unique flavor compounds through metabolic cross-feeding. When certain non-Saccharomyces yeasts die off, they release nutrients that benefit surviving species, leading to the production of novel aromatic compounds not found in single-strain fermentations.

Climate change has begun affecting wine microbiomes globally. Rising temperatures alter the balance of native yeasts on grape surfaces, potentially changing regional wine characteristics. Understanding these shifts becomes crucial for maintaining traditional wine styles while adapting to changing conditions.

Barrel aging introduces additional microbial complexity to wine. The porous nature of oak harbors distinct microbiological communities that slowly influence wine development. These barrel-associated microbes contribute to the development of aged wine characteristics through subtle metabolic activities.

The presence of specific yeast strains can influence grape variety expression. Some indigenous yeasts enhance varietal thiols in Sauvignon Blanc, while others promote the development of spicy notes in Shiraz. This strain-specific impact on varietal character highlights the importance of preserving regional microbial diversity.

Modern winemaking techniques must balance the desire for consistency with the benefits of microbial diversity. Some producers now deliberately work with mixed cultures, combining selected strains with indigenous populations to achieve both reliability and complexity.

The study of wine microbiomes has revealed unexpected connections between seemingly unrelated wines from different regions. Similar microbial patterns can emerge in vineyards thousands of miles apart, suggesting universal principles in how these communities develop and function.

Understanding the wine microbiome helps explain the phenomenon of stuck fermentations. These problematic situations often result from microbial competition and stress responses, rather than simple nutrient depletion as previously thought.

The persistence of certain microbes throughout fermentation can influence wine stability and aging potential. Some species produce compounds that act as natural preservatives, while others contribute to the wine's antioxidant capacity.

As our knowledge of wine microbiomes deepens, new opportunities emerge for enhancing wine quality and expression. By working with rather than against these complex microbial communities, winemakers can better preserve and enhance the unique characteristics that make each wine special.

# Spatial and Temporal Microbial Dynamics

Microbial communities in fermentation systems demonstrate remarkable patterns of organization across both space and time. These dynamic arrangements shape the development of fermented foods and beverages, creating complex networks of interaction that evolve throughout the fermentation process.

The spatial distribution of microorganisms within fermentation vessels follows distinct patterns that optimize resource utilization and metabolic efficiency. In traditional ceramic kimchi containers, for instance, different bacterial species arrange themselves in layers, with aerobic organisms dominating the surface while anaerobic bacteria thrive in deeper regions. This stratification creates distinct microzones that contribute to the final product's complexity.

Temporal succession in fermented foods follows predictable yet intricate patterns. Early colonizers modify the environment through their metabolic activities, creating conditions that favor subsequent organisms. This progression can be observed in sourdough fermentation, where initial acid-tolerant bacteria prepare the environment for specialized yeast species that later dominate the culture.

Physical barriers and surface attachments play crucial roles in microbial organization. Biofilm formation on the surface of fermenting vegetables creates protective microenvironments that shield beneficial organisms from environmental stresses. These structured communities develop sophisticated architectures that facilitate nutrient exchange and communication between different species.

Temperature gradients within fermentation vessels influence microbial distribution patterns. Warmer regions often support more active metabolism, while cooler areas may serve as reservoirs for slower-growing species. This thermal stratification contributes to the development of diverse flavor compounds through varied metabolic activities.

The movement of nutrients and metabolites through fermentation systems creates chemical gradients that guide microbial positioning. Species sensitive to certain compounds will migrate to more favorable zones, while others may cluster around nutrient-rich areas. This self-organization optimizes resource utilization across the entire community.

Population dynamics fluctuate dramatically during different fermentation phases. Initial explosive growth of pioneer species gives way to more stable communities as resources become limited. These transitions often coincide with significant changes in product characteristics, such as the development of complex flavors in aged cheeses.

Microbial interactions across spatial boundaries influence fermentation outcomes. In traditional pit fermentation, microorganisms in the surrounding soil contribute to the process by exchanging metabolites with the fermenting material. These extended networks of interaction add layers of complexity to the final product.

The formation of microcolonies within fermentation systems creates localized centers of activity. These clusters of related organisms work together to process available nutrients while protecting themselves from competition. The spacing between these colonies affects the rate and uniformity of fermentation.

Seasonal variations impact microbial community composition and activity patterns. Traditional fermented foods often exhibit different characteristics depending on the time of year they were produced, reflecting changes in the available microbiota and environmental conditions.

Surface area to volume ratios in fermentation vessels influence microbial distribution and activity. Larger surface areas promote the growth of aerobic organisms, while greater depths support anaerobic processes. Understanding these relationships helps optimize vessel design for desired outcomes.

The movement of liquids within fermentation systems, whether through natural convection or mechanical agitation, affects microbial distribution patterns. This fluid dynamics can create zones of mixing that promote interaction between different species while maintaining distinct microbial territories.

Cell density gradients develop naturally during fermentation, with some areas supporting higher populations than others. These variations in population density influence local metabolic activities and contribute to the development of product complexity.

The attachment of microorganisms to solid surfaces creates stable communities that persist throughout fermentation. These attached populations often serve as reservoirs of genetic diversity, helping maintain consistent characteristics across multiple fermentation cycles.

Time-dependent changes in pH and oxygen availability drive shifts in microbial community composition. Early fermentation stages often see rapid changes in these parameters, leading to successive waves of microbial activity that shape product development.

The formation of metabolic partnerships between different species requires proper spatial arrangement. Organisms that depend on each other's metabolic byproducts tend to position themselves in close proximity, creating efficient exchange networks that benefit the entire community.

## Predictive Modeling of Microbial Behavior

Mathematical models have emerged as powerful tools for understanding and predicting the complex behaviors of microorganisms during fermentation processes. These sophisticated approaches combine biological knowledge with computational methods to forecast how microbial communities will develop and interact under various conditions.

Growth kinetics form the foundation of predictive modeling, incorporating factors such as temperature, pH, water activity, and nutrient availability. By understanding how these parameters influence microbial growth rates, researchers can develop accurate predictions for fermentation outcomes. For instance, the classic Monod equation has been adapted to account for the multiple substrate limitations often encountered in real fermentation systems.

Environmental stress responses play a crucial role in model development. Microorganisms react to changing conditions in ways that affect their metabolism and growth patterns. Modern predictive models incorporate these stress responses, allowing for more accurate forecasting of fermentation trajectories under varying conditions.

Competition between different species significantly influences fermentation dynamics. Mathematical models now account for these interactions, considering how various organisms compete for resources while potentially producing compounds that inhibit or promote the growth of other species. This interplay creates complex feedback loops that shape the final product characteristics.

Temperature fluctuations during fermentation can be modeled using differential equations that describe heat transfer and metabolic activity. These models help predict how temperature changes affect microbial growth rates and metabolite production, enabling better control of fermentation processes.

The succession of different microbial species throughout fermentation follows patterns that can be predicted through careful modeling. Early colonizers modify the environment in ways that influence which organisms will thrive in subsequent stages. Understanding these succession patterns helps optimize fermentation timing and conditions.

Metabolic flux analysis provides insights into how nutrients flow through microbial communities. By modeling these fluxes, researchers can predict the formation of important flavor compounds and other metabolites that contribute to product quality.

Stochastic elements in microbial behavior require special consideration in predictive modeling. Random variations in growth rates and metabolic activities can significantly impact fermentation outcomes. Modern models incorporate these probabilistic aspects to provide more realistic predictions.

The spatial distribution of microorganisms within fermentation vessels affects local conditions and interaction patterns. Three-dimensional models can predict how different species will organize themselves based on environmental gradients and metabolic requirements.

Time-series analysis helps identify critical points in fermentation processes where specific interventions might be necessary. These models can predict when conditions might become unfavorable for desired organisms or when harmful compounds might accumulate to dangerous levels.

Population dynamics models reveal how different species interact over time. These interactions can be cooperative or competitive, and understanding them helps predict how community composition will evolve throughout fermentation.

pH changes during fermentation follow predictable patterns that can be modeled mathematically. These models account for acid production by various organisms and the buffering capacity of the substrate, helping maintain optimal conditions for desired outcomes.

Substrate utilization patterns provide important information for predictive modeling. Different microorganisms prefer specific nutrients and produce various metabolites, creating complex networks of resource use that can be mapped and predicted.

Machine learning approaches have enhanced our ability to predict microbial behavior by identifying subtle patterns in large datasets. These models can account for numerous variables simultaneously, providing more accurate predictions than traditional statistical methods alone.

The formation of beneficial compounds during fermentation follows kinetic patterns that can be modeled mathematically. Understanding these patterns helps optimize conditions for maximum production of desired metabolites while minimizing unwanted byproducts.

Oxygen availability significantly influences microbial behavior and can be modeled using mass transfer equations. These models help predict how different aeration levels will affect microbial growth and metabolism throughout the fermentation process.

Environmental fluctuations in industrial settings can be incorporated into predictive models to account for real-world variations. This includes considerations of equipment limitations and operational constraints that might affect fermentation outcomes.

The integration of multiple modeling approaches provides the most comprehensive understanding of microbial behavior. Combining metabolic models with population dynamics and environmental parameters creates powerful predictive tools for fermentation control.

As our understanding of microbial genetics improves, predictive models increasingly incorporate genomic information to forecast how different strains will behave under various conditions. This integration of molecular data with traditional modeling approaches enhances prediction accuracy and reliability.

These sophisticated modeling techniques continue to evolve, providing increasingly accurate predictions of microbial behavior in fermentation systems. Their application helps bridge the gap between traditional fermentation practices and modern scientific understanding, enabling better control and optimization of fermentation processes.

## Bioinformatic Tools for Microbial Exploration

Modern bioinformatic approaches have revolutionized our understanding of microbial communities in fermented foods, offering unprecedented insights into their composition, function, and interactions. These computational tools transform raw genetic data into meaningful information about the microorganisms that drive fermentation processes.

Sequence analysis software forms the foundation of microbial community studies. These tools process millions of DNA sequences from fermented foods, identifying individual species and strains present in complex mixtures. Advanced algorithms can distinguish between closely related organisms, revealing the true diversity of fermentation microbiomes.

Taxonomic classification tools help researchers organize microbial data into meaningful hierarchies. By comparing genetic sequences against established databases, these programs identify both known and novel organisms present in fermented foods. This classification process reveals previously unknown players in traditional fermentations.

Metagenome assembly software pieces together fragmentary DNA sequences to reconstruct complete microbial genomes. This process, like solving an enormous puzzle, reveals the genetic potential of entire communities. Understanding these genetic blueprints helps predict how different organisms might contribute to fermentation outcomes.

Metabolic pathway analysis tools map the biochemical capabilities of fermentation microorganisms. By examining gene content, researchers can predict which compounds different species might produce or consume during fermentation. This information helps optimize conditions for desired flavor development.

Comparative genomics software identifies differences between strains of the same species. These subtle variations often explain why certain strains perform better in specific fermentation contexts. Understanding these differences helps in selecting optimal starter cultures for different products.

Network analysis tools reveal complex interactions between different microorganisms in fermented foods. These programs construct visual representations of how various species influence each other through metabolic exchanges, competition, and cooperation.

Statistical analysis packages help researchers interpret complex datasets from microbial studies. These tools identify significant patterns and correlations that might otherwise remain hidden in the vast amount of data generated by modern sequencing techniques.

Visualization software transforms complex microbial data into comprehensible graphics. These tools help researchers communicate their findings effectively and identify important patterns in community composition and function.

Database management systems organize and store vast amounts of microbial information. These repositories allow researchers to compare their findings with previous studies and identify common patterns across different fermented foods.

Quality control tools ensure the reliability of genetic data by filtering out sequencing errors and contamination. This crucial step prevents misleading conclusions about microbial community composition and function.

Functional annotation software predicts the roles of different genes identified in fermentation microorganisms. These predictions help researchers understand how various species might contribute to flavor development and product characteristics.

Phylogenetic analysis tools reconstruct evolutionary relationships between different microorganisms. This information helps track the spread of beneficial traits through microbial populations and understand how different strains adapt to specific fermentation environments.

Time-series analysis software tracks changes in microbial communities throughout the fermentation process. These tools reveal succession patterns and help identify critical points where community composition shifts significantly.

Machine learning applications identify subtle patterns in microbial data that might escape traditional analysis methods. These tools can predict fermentation outcomes based on initial community composition and environmental conditions.

Population genetics software examines how microbial populations change over time and space. These tools help researchers understand how different strains spread through fermentation facilities and adapt to specific conditions.

Primer design tools help researchers develop specific tests for important microorganisms. These programs ensure that genetic markers accurately identify target species while avoiding cross-reaction with similar organisms.

Data integration platforms combine information from multiple sources to provide comprehensive views of microbial communities. These tools help researchers connect genetic data with metabolic measurements and product characteristics.

Version control systems track changes in analysis procedures and results over time. This documentation ensures reproducibility and allows researchers to refine their methods as new information becomes available.

Pipeline management tools automate complex sequences of analysis steps. These workflows ensure consistent processing of microbial data across different samples and studies.

The rapid evolution of bioinformatic tools continues to expand our understanding of fermentation microbiomes. Each new development reveals additional layers of complexity in these fascinating microbial communities, while simultaneously making their study more accessible and comprehensive.

These computational approaches bridge the gap between traditional fermentation knowledge and modern scientific understanding. By revealing the intricate details of microbial communities, bioinformatic tools help preserve and improve traditional fermented foods while developing new products that harness the power of beneficial microorganisms.

# Chapter 4: Engineering Microbial Communities

## Probiotics and Synbiotics in Winemaking

The integration of probiotics and synbiotics into winemaking represents an innovative approach to enhancing both the fermentation process and the potential health benefits of wine. Traditional winemaking practices have inadvertently incorporated beneficial microorganisms for centuries, but modern understanding allows for their deliberate and optimized use.

Probiotic strains selected for winemaking must withstand challenging conditions, including high alcohol content, low pH, and the presence of sulfites. Specific strains of Lactobacillus and Pediococcus have shown remarkable resilience in wine environments, contributing positively to both fermentation dynamics and potential health benefits.

The selection of appropriate probiotic strains begins with careful screening for oenological compatibility. These microorganisms must not only survive but also contribute positively to wine characteristics without producing off-flavors or unwanted compounds. Successful strains often demonstrate abilities to enhance color stability, improve mouthfeel, and contribute to the wine's aromatic complexity.

Synbiotic relationships in wine develop through the careful combination of probiotic organisms with prebiotics naturally present in grape must. These relationships can be enhanced by selecting grape varieties with higher concentrations of specific polyphenols and oligosaccharides that support probiotic growth and survival.

Timing of probiotic introduction proves crucial for successful integration. Some winemakers add probiotic strains during primary fermentation, while others prefer introduction during malolactic fermentation when conditions become more favorable for these beneficial bacteria. Each approach offers distinct advantages and challenges that must be carefully considered.

The interaction between probiotic bacteria and wine yeasts creates complex relationships that influence fermentation outcomes. Some probiotic strains demonstrate positive synergistic effects with Saccharomyces cerevisiae, leading to more complete fermentations and enhanced flavor development. However, careful strain selection remains essential to avoid antagonistic interactions.

Temperature control becomes particularly critical when working with probiotics in wine. Most beneficial strains show optimal activity between 20-25°C, requiring careful management of fermentation temperatures to maintain viability while ensuring proper wine development.

The presence of sulfites, commonly used as preservatives in winemaking, can challenge probiotic survival. Modern approaches include selecting sulfite-resistant strains and developing alternative preservation methods that maintain both probiotic viability and wine stability.

Monitoring probiotic populations throughout the winemaking process requires specialized techniques. Regular viability checks help ensure that beneficial organisms maintain sufficient populations to confer their intended benefits. Modern molecular methods allow precise tracking of specific strains throughout fermentation.

The impact of probiotics on wine aging characteristics presents both opportunities and challenges. Some strains contribute to improved color stability and enhanced complexity during aging, while others may influence the development of tertiary aromas in unexpected ways.

Storage conditions significantly affect probiotic survival in finished wines. Temperature fluctuations, oxygen exposure, and light can all impact the viability of beneficial organisms. Proper storage protocols help maintain both wine quality and probiotic benefits.

The development of synbiotic relationships extends beyond simple probiotic-prebiotic interactions. Complex networks form between different microorganisms, grape compounds, and environmental factors, creating unique ecological niches within the wine matrix.

Consumer acceptance of probiotic wines continues to grow as understanding of their potential benefits increases. Educational efforts focus on explaining how these wines differ from traditional products while maintaining familiar quality characteristics.

Quality control measures for probiotic wines require additional considerations beyond standard wine analysis. Regular monitoring of probiotic viability, metabolic activity, and potential impact on wine characteristics ensures consistent product quality.

The legal framework surrounding probiotic wine production varies by region. Some areas have established specific guidelines for probiotic wine labeling and production, while others continue to develop appropriate regulations.

Research into new probiotic strains specifically adapted for winemaking continues to expand possibilities. Novel strains isolated from wine environments often show superior adaptation to wine conditions while contributing unique beneficial characteristics.

The integration of traditional winemaking knowledge with modern probiotic science creates opportunities for innovation while respecting established practices. This balance helps maintain wine quality while enhancing potential health benefits.

Economic considerations play a significant role in probiotic wine production. While these wines may command premium prices, additional production costs and monitoring requirements must be carefully balanced against market potential.

The future of probiotic winemaking lies in understanding and optimizing the complex interactions between beneficial microorganisms and traditional wine characteristics. Continued research and development promise new approaches to enhancing both the quality and potential health benefits of wine through careful microbial management.

Success in probiotic winemaking requires a holistic approach that considers all aspects of production, from grape selection through final packaging. This comprehensive strategy ensures that both traditional wine quality and probiotic benefits are maintained throughout the entire process.

## Microbial Consortia and Engineered Symbiosis

The intricate world of microbial consortia represents nature's most sophisticated example of biological cooperation and competition. These complex communities of microorganisms work together in ways that enhance their collective capabilities far beyond what any single species could achieve alone.

Natural microbial consortia have evolved over millions of years to perform complex biochemical transformations efficiently. In fermentation processes, these communities demonstrate remarkable stability and resilience, adapting to environmental changes while maintaining essential functions. Understanding these natural partnerships provides the foundation for engineering improved symbiotic relationships.

The development of engineered symbiotic relationships begins with careful selection of compatible microorganisms. Each member of the consortium must contribute unique capabilities while avoiding harmful competition with other members. This delicate balance requires deep understanding of metabolic networks and interaction patterns.

Cross-feeding relationships form the backbone of successful microbial consortia. One organism's waste products become essential nutrients for another, creating efficient metabolic cycles that minimize resource waste. These relationships can be enhanced through careful strain selection and environmental optimization.

Spatial organization plays a crucial role in consortium function. Microorganisms arrange themselves in complex three-dimensional structures that facilitate nutrient exchange and communication. Engineering these spatial relationships can improve consortium stability and productivity.

Communication between consortium members occurs through various chemical signals. These molecular messages coordinate group behaviors and regulate metabolic activities. Understanding and optimizing these communication networks helps create more effective engineered systems.

Stability mechanisms in natural consortia provide valuable lessons for engineering robust systems. Redundant metabolic pathways and multiple species performing similar functions help maintain community function even when individual members face challenges.

The selection of anchor species forms a critical step in consortium design. These key organisms provide essential functions or create environmental conditions that support other consortium members. Their careful choice can determine the success or failure of the entire community.

Environmental factors significantly influence consortium behavior. Temperature, pH, nutrient availability, and oxygen levels must be carefully controlled to maintain optimal performance. These parameters often require dynamic adjustment as the community develops and changes over time.

Metabolic modeling helps predict how different species will interact within a consortium. These predictions guide strain selection and process optimization, reducing the time and resources required for successful consortium development.

Stress responses within consortia often differ from those observed in pure cultures. Community members can protect each other from environmental challenges, leading to increased resilience. Engineering these protective relationships enhances consortium stability.

The development of synthetic consortia offers opportunities to combine beneficial traits from different sources. Carefully selected combinations of wild-type and engineered strains can create communities with enhanced capabilities for specific applications.

Population dynamics within consortia follow complex patterns that change over time. Understanding these patterns helps maintain desired community compositions through appropriate process control strategies.

Genetic stability presents particular challenges in engineered consortia. Mechanisms to prevent the loss of desired traits or the emergence of detrimental mutations must be incorporated into consortium design.

Resource competition between consortium members requires careful management. Strategic limitation of specific nutrients can help maintain desired population ratios and prevent individual species from dominating the community.

Monitoring consortium composition and function presents unique challenges. Advanced analytical techniques allow tracking of individual species while measuring important metabolic products and intermediates.

The scale-up of successful consortia from laboratory to industrial scale requires careful attention to maintaining appropriate conditions. Factors that work well in small-scale systems may need modification for larger operations.

Product consistency from consortium-based processes depends on maintaining stable community composition. Regular monitoring and adjustment of operating conditions helps ensure reliable performance over extended periods.

Regulatory considerations for engineered consortia often differ from those for single-strain processes. Safety assessments must consider potential interactions between community members and their combined effects on product quality.

The evolution of consortia over time can lead to both improved performance and potential problems. Regular assessment of community composition and function helps identify beneficial changes while catching potential issues early.

The future of engineered microbial consortia lies in creating increasingly sophisticated communities capable of performing complex transformations efficiently and reliably. These advances will continue to expand the possibilities for sustainable bioprocessing and fermentation technologies.

Success in developing effective microbial consortia requires combining knowledge from multiple disciplines, including microbiology, biochemistry, genetics, and process engineering. This integrated approach leads to robust and productive communities that advance the field of industrial fermentation.

## Selective Pressures and Microbial Adaptation

Microorganisms in fermentation environments face constant evolutionary challenges that shape their genetic makeup and metabolic capabilities. These selective pressures drive adaptation processes, resulting in specialized strains uniquely suited to specific fermentation conditions.

Environmental stressors in fermentation systems create powerful selective forces. High alcohol concentrations, extreme pH levels, temperature fluctuations, and nutrient limitations all contribute to evolutionary pressure. Successful strains must develop mechanisms to cope with these challenges while maintaining their essential metabolic functions.

The adaptation process begins with existing genetic variation within microbial populations. Some individuals naturally possess traits that provide advantages under specific conditions. These beneficial characteristics become more prevalent in subsequent generations as better-adapted organisms outcompete their less successful counterparts.

Horizontal gene transfer plays a crucial role in microbial adaptation during fermentation. Genetic material exchanged between different species can provide immediate access to beneficial traits, accelerating the adaptation process. This mechanism explains how resistance to various stressors can spread rapidly through microbial communities.

Competition for resources drives significant evolutionary changes in fermentation microorganisms. Strains that can efficiently utilize available nutrients while producing compounds that inhibit competitors gain significant advantages. This competitive pressure often leads to the development of unique metabolic pathways and defense mechanisms.

Temperature adaptation represents one of the most significant selective pressures in fermentation environments. Microorganisms must adjust their cellular machinery to function efficiently at specific temperature ranges. This adaptation affects everything from membrane composition to enzyme stability.

pH tolerance mechanisms evolve through constant exposure to acidic or alkaline conditions. Successful strains develop enhanced cellular transport systems and modified cell wall structures to maintain internal pH balance. These adaptations often become permanently encoded in the genome after multiple generations.

Alcohol tolerance emerges as a critical adaptive trait in many fermentation processes. Microorganisms must develop mechanisms to maintain membrane integrity and cellular function despite increasing ethanol concentrations. This adaptation often involves changes in membrane lipid composition and stress response pathways.

Nutrient limitation drives the evolution of more efficient metabolic pathways. Strains that can maximize resource utilization while minimizing energy expenditure gain significant advantages. This optimization often leads to streamlined metabolic networks and enhanced substrate specificity.

Oxidative stress resistance develops as microorganisms cope with varying oxygen levels during fermentation. Adapted strains possess enhanced antioxidant systems and modified metabolic pathways that function effectively under different oxygen concentrations.

The speed of adaptation varies significantly depending on generation time and selection intensity. Rapid reproduction rates in many fermentation microorganisms allow for quick adaptation to new conditions, sometimes occurring within days or weeks rather than years.

Population bottlenecks during fermentation processes can accelerate adaptation by reducing genetic diversity and allowing beneficial mutations to spread rapidly. However, these same bottlenecks can also lead to the loss of potentially useful traits.

Cross-protection mechanisms often develop as microorganisms adapt to multiple stressors simultaneously. Adaptation to one type of stress frequently provides protection against other challenges, creating more robust and versatile strains.

The stability of adapted traits varies depending on the underlying genetic mechanisms. Some adaptations become permanently fixed in the genome, while others may be lost if selective pressure is removed. Understanding these differences helps maintain desired characteristics in production strains.

Biofilm formation represents an important adaptive strategy in many fermentation environments. The ability to form and maintain biofilms provides protection against environmental stressors while facilitating resource sharing within microbial communities.

The evolution of metabolic cooperation between different species creates complex adaptive networks. Strains that can effectively participate in these networks often show enhanced survival and productivity compared to isolated populations.

Selection for desired product characteristics often occurs alongside adaptation to environmental stressors. Strains that maintain or enhance valuable metabolic pathways while developing stress resistance become particularly valuable for industrial applications.

The role of mobile genetic elements in adaptation cannot be understated. Plasmids, transposons, and other mobile elements facilitate rapid acquisition of beneficial traits, allowing microorganisms to quickly respond to new challenges.

Modern techniques for monitoring adaptation processes provide unprecedented insights into evolutionary mechanisms. These tools help identify key genetic changes and metabolic adjustments that contribute to improved strain performance.

The practical implications of microbial adaptation extend beyond individual strains to entire fermentation processes. Understanding these adaptive mechanisms helps optimize conditions for desired outcomes while preventing the development of problematic characteristics.

Success in managing selective pressures requires balancing the need for adaptation with maintaining desired strain characteristics. This delicate equilibrium determines the long-term success of fermentation processes and the quality of final products.

# Epigenetic Regulation of Microbial Traits

Epigenetic modifications represent a fascinating layer of control over microbial behavior in fermentation processes, extending far beyond the basic genetic code. These heritable changes in gene expression occur without alterations to the DNA sequence itself, providing microorganisms with remarkable flexibility in responding to environmental challenges.

DNA methylation patterns play a crucial role in regulating gene expression during fermentation. These chemical modifications can activate or silence specific genes, allowing microorganisms to fine-tune their metabolic responses. The pattern of methylation often reflects the environmental conditions experienced by previous generations, creating a form of cellular memory.

Histone-like proteins in bacteria and other microorganisms undergo modifications that affect DNA accessibility and gene expression. These changes can occur rapidly in response to environmental signals, allowing quick adaptation to changing fermentation conditions. The modifications persist through multiple generations, influencing the behavior of descendant cells.

Small RNA molecules serve as important epigenetic regulators in fermentation microorganisms. These molecules can target specific genes, altering their expression patterns without changing the underlying DNA sequence. The production and activity of these regulatory RNAs often respond to environmental stresses common in fermentation processes.

Stress responses trigger complex epigenetic changes that can persist long after the initial stimulus has passed. Microorganisms exposed to heat, acid, or alcohol stress often maintain modified gene expression patterns that enhance their resistance to future challenges. This acquired stress tolerance can significantly impact fermentation performance.

The inheritance of epigenetic modifications creates populations with enhanced fitness for specific fermentation conditions. Offspring cells inherit not only genetic material but also the beneficial expression patterns established in parent cells. This non-genetic inheritance can rapidly spread advantageous traits through a population.

Environmental factors during fermentation directly influence epigenetic states. Temperature, pH, nutrient availability, and oxygen levels can all trigger specific modifications that alter microbial behavior. Understanding these relationships helps optimize fermentation conditions for desired outcomes.

Metabolic regulation through epigenetic mechanisms allows microorganisms to efficiently allocate resources. Genes involved in unnecessary pathways can be temporarily silenced, while those required for current conditions are activated. This flexibility helps maintain energy efficiency during fermentation.

The stability of epigenetic modifications varies significantly among different traits and conditions. Some changes persist for many generations, while others revert quickly when conditions change. This variable stability influences the long-term behavior of fermentation cultures.

Competition between microorganisms can trigger epigenetic changes that enhance survival capabilities. Exposure to competing strains often leads to modified expression patterns for genes involved in resource acquisition and defense mechanisms. These adaptations can significantly impact community dynamics in mixed fermentations.

The timing of epigenetic modifications proves crucial for successful fermentation outcomes. Early exposure to specific conditions can establish beneficial expression patterns that persist throughout the process. This temporal aspect must be considered when designing fermentation protocols.

Nutrient availability strongly influences epigenetic regulation patterns. Microorganisms adjust their metabolism through epigenetic modifications to optimize resource utilization. These adjustments can affect both growth rates and product formation.

The transmission of epigenetic information between different species occurs through various molecular signals. This cross-talk can coordinate behavior within microbial communities, leading to more stable and productive fermentations.

Industrial applications of epigenetic regulation include the development of improved fermentation strains. Controlled exposure to specific conditions can establish beneficial expression patterns that enhance productivity or stress tolerance. These improvements often remain stable through many generations of industrial use.

Modern analytical techniques allow detailed monitoring of epigenetic states during fermentation. This information helps identify critical control points where environmental conditions can be adjusted to maintain desired expression patterns.

The role of epigenetics in strain evolution continues to reveal new insights into adaptation mechanisms. Understanding how expression patterns change over time helps predict and control the long-term behavior of fermentation cultures.

Quality control in industrial fermentations must consider epigenetic stability. Regular monitoring of key traits helps ensure that beneficial modifications persist through multiple generations of culture propagation.

The interaction between genetic and epigenetic regulation creates complex networks of control over microbial behavior. These networks provide multiple layers of adaptation capability, enhancing the robustness of fermentation processes.

Future developments in understanding epigenetic regulation will likely reveal new opportunities for process optimization. The ability to deliberately manipulate expression patterns without genetic modification offers promising approaches for strain improvement.

Success in managing epigenetic regulation requires careful attention to environmental conditions throughout the fermentation process. Maintaining consistent conditions helps preserve beneficial modifications while preventing unwanted changes in expression patterns.

## Microbial Interventions for Flavor and Quality

The intricate relationship between microorganisms and flavor development stands at the heart of fermented food and beverage production. Through carefully selected microbial interventions, producers can significantly enhance product quality while creating unique and desirable flavor profiles that distinguish their offerings in the marketplace.

Lactic acid bacteria play a fundamental role in developing complex flavor compounds through their metabolic activities. These industrious microorganisms break down proteins into peptides and free amino acids, creating precursor molecules that contribute to the final product's taste and aroma. Their ability to produce various organic acids also adds depth to flavor profiles while contributing to preservation.

Yeast strains contribute an array of flavor-active compounds through their fermentation pathways. Higher alcohols, esters, and phenolic compounds emerge from their metabolism, creating distinctive aromatic signatures in products ranging from wine to sourdough bread. The selection of specific yeast strains can dramatically influence the final sensory characteristics.

The timing of microbial additions significantly impacts flavor development. Sequential inoculation strategies allow different organisms to contribute their unique characteristics at optimal moments during fermentation. This careful orchestration of microbial activity helps achieve complex and balanced flavor profiles.

Temperature control during fermentation directly affects microbial metabolism and subsequent flavor production. Lower temperatures often result in cleaner, more delicate flavors, while higher temperatures can promote the formation of bold, intense flavor compounds. Understanding these relationships enables precise control over product characteristics.

Nutrient availability influences the production of flavor-active compounds by fermentation microorganisms. Careful manipulation of nitrogen sources, minerals, and vitamins can guide metabolic pathways toward desired flavor outcomes. This nutritional management becomes particularly important in large-scale production settings.

The interaction between different microbial species creates unique flavor combinations impossible to achieve with single-strain fermentations. These synergistic relationships often produce complex flavor profiles that exceed the sum of individual contributions. Careful selection of compatible strains becomes crucial for successful outcomes.

Proteolytic activities of specific microorganisms release flavor precursors from protein-rich substrates. This breakdown process generates peptides and amino acids that contribute directly to taste while serving as building blocks for additional flavor compounds. The extent of proteolysis must be carefully controlled to achieve desired results.

Oxidative processes managed by certain microorganisms can enhance or deteriorate product quality. While some oxidation may contribute positively to flavor development, excessive activity can lead to off-flavors and quality defects. Monitoring and controlling these processes ensures consistent product quality.

The formation of bioactive compounds through microbial activity often correlates with flavor development. Many of these compounds contribute not only to sensory characteristics but also to the nutritional and functional properties of fermented products. This dual benefit adds value to microbial interventions.

Strain selection criteria must balance flavor production capabilities with technological properties. Microorganisms need to perform reliably under production conditions while consistently generating desired flavor compounds. This balance often requires extensive screening and testing programs.

The management of off-flavor formation presents ongoing challenges in fermentation processes. Certain microbial activities can produce undesirable compounds that detract from product quality. Prevention strategies often focus on controlling growth conditions and selecting appropriate starter cultures.

Environmental stress responses in microorganisms can significantly impact flavor production. Understanding how various stressors affect metabolic pathways allows producers to manipulate conditions for optimal flavor development. This knowledge proves particularly valuable when scaling up production processes.

The preservation of desired flavor characteristics requires careful attention to storage conditions after fermentation. Some flavor compounds continue to evolve during aging or storage, while others may degrade. Managing these changes ensures product quality throughout the shelf life.

Modern analytical techniques enable precise monitoring of flavor compound production during fermentation. This capability allows for real-time adjustments to process parameters, helping maintain consistency in large-scale operations. Regular analysis also helps identify potential quality issues early in production.

The development of new starter cultures focuses increasingly on enhanced flavor production capabilities. Strain improvement programs consider both traditional breeding approaches and modern selection techniques to create microorganisms with superior flavor-producing characteristics.

Quality control procedures must account for the complex nature of microbial flavor production. Regular sensory evaluation combined with chemical analysis helps maintain product consistency while identifying potential improvements in production processes.

Success in microbial flavor development requires deep understanding of both the organisms involved and the chemical pathways leading to desired compounds. This knowledge, combined with careful process control, enables the consistent production of high-quality fermented products with distinctive flavor profiles.

Traditional fermentation practices often provide valuable insights into effective microbial interventions for flavor development. Many time-tested techniques reveal sophisticated understanding of microbial behavior, offering lessons that remain relevant in modern production settings.

The future of microbial flavor production lies in combining traditional knowledge with modern scientific understanding. This synthesis enables the development of innovative products while maintaining the authentic characteristics that consumers value in fermented foods and beverages.

# Chapter 5: The Quantum Leap in Wine Production

## Quantum Computing and Microbial Optimization

Quantum computing represents a revolutionary approach to solving complex microbial optimization problems in fermentation processes. The unique capabilities of quantum systems offer unprecedented potential for modeling and predicting microbial behavior at molecular and population levels.

The application of quantum algorithms to metabolic pathway analysis reveals intricate patterns previously hidden from traditional computational methods. These insights enable deeper understanding of how microorganisms respond to environmental changes during fermentation. By processing multiple potential outcomes simultaneously, quantum computing helps identify optimal conditions for desired metabolic activities.

Quantum-assisted modeling of protein folding dynamics provides crucial information about enzyme behavior under various fermentation conditions. This advanced computational capability allows researchers to predict how structural changes affect catalytic efficiency, leading to more precise control over fermentation processes. The ability to simulate protein interactions at the quantum level reveals subtle mechanisms that influence microbial performance.

The optimization of microbial communities benefits significantly from quantum computing's ability to process complex interaction networks. By analyzing countless possible combinations of species and environmental conditions simultaneously, these systems help identify ideal community compositions for specific fermentation goals. The quantum advantage becomes particularly evident when dealing with large-scale industrial processes involving multiple bacterial strains.

Environmental parameter optimization through quantum algorithms enables more efficient process development. These sophisticated computational tools can rapidly evaluate numerous combinations of temperature, pH, nutrient concentrations, and other critical variables. The result is faster identification of optimal conditions for maximum productivity and product quality.

Strain development programs leverage quantum computing to predict genetic modifications most likely to yield desired improvements. By simulating multiple genetic variations simultaneously, researchers can identify promising combinations that traditional computing approaches might miss. This capability accelerates the development of enhanced production strains.

The analysis of metabolic flux distributions benefits from quantum computing's unique ability to handle complex probability calculations. Understanding how nutrients flow through various metabolic pathways helps optimize feed strategies and improve product yields. Quantum algorithms excel at processing the massive datasets generated by metabolomic studies.

Real-time process control systems enhanced by quantum computing enable more responsive and precise fermentation management. These advanced systems can predict process outcomes and suggest adjustments before problems develop. The ability to process multiple sensor inputs simultaneously leads to more stable and efficient operations.

Quality control procedures become more sophisticated through quantum-assisted pattern recognition. These systems can identify subtle variations in product characteristics that might indicate process drift or contamination. Early detection of potential issues allows for prompt corrective action.

The optimization of scale-up procedures benefits from quantum computing's ability to model complex fluid dynamics and mass transfer phenomena. These calculations help predict how changes in vessel size and geometry will affect microbial performance. More accurate scaling predictions lead to smoother transitions from laboratory to production scale.

Energy efficiency improvements in fermentation processes emerge from quantum-optimized operating parameters. By analyzing multiple variables simultaneously, these systems identify conditions that maximize productivity while minimizing energy consumption. This capability becomes increasingly important as industries focus on sustainability.

Product recovery and purification strategies benefit from quantum-assisted process modeling. These advanced computational tools help optimize separation parameters and predict product quality outcomes. The ability to simulate multiple separation scenarios simultaneously leads to more efficient downstream processing.

The development of new fermentation products accelerates through quantum-enhanced molecular modeling. These systems can predict how different compounds will interact with microbial metabolism, speeding the discovery of novel products. The quantum advantage becomes particularly valuable when exploring large chemical space.

Storage stability predictions improve through quantum-assisted modeling of product degradation pathways. These calculations help identify optimal storage conditions and predict shelf life more accurately. Understanding molecular-level changes during storage leads to better preservation strategies.

Cost optimization benefits from quantum computing's ability to process complex economic models alongside biological parameters. These systems can identify the most cost-effective operating conditions while maintaining product quality. The simultaneous evaluation of multiple variables leads to more profitable operations.

Regulatory compliance becomes more manageable through quantum-enhanced process validation. These systems help identify critical control points and predict potential compliance issues before they occur.

More comprehensive process understanding leads to more robust quality assurance programs.

The integration of quantum computing with traditional bioprocess control systems creates powerful new tools for fermentation optimization. These hybrid approaches combine the best aspects of both technologies, leading to more effective process management strategies.

Future developments in quantum computing will likely reveal new opportunities for understanding and controlling microbial behavior. As quantum systems become more powerful and accessible, their impact on fermentation technology will continue to grow.

Success in applying quantum computing to microbial optimization requires close collaboration between computational experts and fermentation specialists. This interdisciplinary approach ensures that technological capabilities align with practical industry needs.

The transformation of fermentation technology through quantum computing represents a significant step forward in process optimization. These advanced computational tools enable more precise control and better understanding of microbial systems, leading to improved products and more efficient processes.

## Quantum Sensors and Real Time Monitoring

Quantum sensors represent a revolutionary advancement in fermentation monitoring, offering unprecedented precision and real-time insight into microbial processes. These sophisticated devices utilize quantum mechanical principles to detect and measure subtle changes at the molecular level, providing crucial data for process optimization and control.

Nitrogen-vacancy (NV) centers in diamond-based quantum sensors enable ultra-sensitive detection of magnetic fields produced by metabolically active microorganisms. These measurements reveal detailed information about cellular activity and metabolism without disturbing the fermentation process. The ability to monitor biological processes non-invasively represents a significant advantage over traditional sampling methods.

Quantum tunneling sensors provide precise measurements of dissolved oxygen levels in fermentation media. These devices detect oxygen molecules with exceptional accuracy, allowing for better control of aerobic and anaerobic conditions. The rapid response time of quantum sensors enables immediate adjustments to maintain optimal oxygen levels throughout the fermentation process.

The application of quantum-enhanced pH monitoring systems delivers unprecedented accuracy in measuring hydrogen ion concentrations. These sensors utilize quantum effects to detect minute changes in pH, crucial for maintaining optimal conditions for microbial growth and product formation. Real-time pH data helps prevent unwanted metabolic shifts that could affect product quality.

Temperature monitoring achieves new levels of precision through quantum thermometry. These advanced sensors can detect temperature variations at the microscale, providing detailed information about heat distribution within fermentation vessels. Understanding thermal gradients helps optimize mixing strategies and heat transfer systems.

Biomass concentration measurements benefit from quantum optical sensors that can penetrate dense cultures with high accuracy. These devices use quantum-enhanced light scattering techniques to determine cell density in real-time, enabling better control of growth rates and substrate feeding strategies.

Product formation monitoring reaches new levels of sophistication through quantum-enhanced spectroscopic methods. These sensors can detect specific molecules with extraordinary sensitivity, allowing continuous tracking of product accumulation. Real-time product data helps optimize harvest timing and process parameters.

Substrate utilization patterns become clearer through quantum sensor arrays that simultaneously monitor multiple components. These systems provide detailed information about metabolic flux, helping optimize feed rates and substrate composition. The ability to track multiple parameters simultaneously improves process understanding and control.

Contamination detection benefits from quantum-enhanced biosensors that can identify foreign organisms at extremely low concentrations. These systems provide early warning of potential contamination events, allowing rapid intervention to protect product quality. The high sensitivity of quantum sensors enables detection before contamination becomes visible through traditional methods.

The integration of quantum sensors with existing control systems creates powerful new capabilities for process automation. Real-time data from multiple quantum sensors enables sophisticated control algorithms to maintain optimal conditions continuously. This integration leads to more stable and efficient fermentation processes.

Metabolic state analysis becomes more precise through quantum-enhanced fluorescence measurements. These sensors can detect subtle changes in cellular fluorescence that indicate shifts in metabolic activity. Understanding these changes helps maintain desired metabolic pathways throughout fermentation.

The monitoring of protein folding and aggregation benefits from quantum-enhanced light scattering techniques. These measurements provide crucial information about product quality in real-time, particularly important for biological products. Early detection of aggregation helps prevent product loss and quality issues.

Viscosity measurements achieve new levels of accuracy through quantum mechanical sensors. These devices can detect subtle changes in fluid properties that might indicate problems with mixing or mass transfer. Continuous viscosity monitoring helps maintain optimal mixing conditions throughout fermentation.

Gas composition analysis benefits from quantum-enhanced spectroscopic methods that can detect multiple components simultaneously. These measurements provide detailed information about metabolic activity and process efficiency. Real-time gas analysis helps optimize aeration and ventilation strategies.

The development of miniaturized quantum sensors enables more comprehensive monitoring of large-scale fermentations. These devices can be distributed throughout production vessels, providing detailed spatial information about process conditions. Better understanding of spatial variations leads to improved mixing and control strategies.

Data integration from multiple quantum sensors creates comprehensive process fingerprints that help identify optimal operating conditions. These detailed datasets enable better understanding of process dynamics and more effective control strategies. The ability to correlate multiple parameters simultaneously leads to better process optimization.

Cost considerations for quantum sensor implementation continue to improve as technology advances. While initial investment may be significant, the benefits of improved process control and reduced product loss often justify the expense. The long-term reliability of quantum sensors contributes to their economic viability.

Future developments in quantum sensing technology will likely reveal new opportunities for process monitoring and control. As sensors become more sophisticated and affordable, their application in fermentation processes will continue to expand. The integration of quantum sensing with other advanced technologies promises even greater improvements in process understanding and control.

Success in implementing quantum sensors requires careful attention to calibration and maintenance procedures. Regular verification of sensor performance ensures reliable data for process control. The development of robust calibration protocols remains crucial for widespread adoption of quantum sensing technology.

The transformation of fermentation monitoring through quantum sensors represents a significant advance in bioprocess technology. These sophisticated devices enable unprecedented insight into microbial processes, leading to better control and more consistent product quality.

## Quantum Teleportation and Microbial Communication

The fascinating intersection of quantum teleportation principles and microbial communication networks reveals unprecedented opportunities for understanding and manipulating cellular interactions. At the molecular level, quantum effects play subtle yet significant roles in how microorganisms exchange information and coordinate their activities during fermentation processes.

Quantum entanglement phenomena have been observed in biological systems, particularly in the electron transport chains of metabolically active microorganisms. These quantum-level interactions contribute to the efficiency of energy transfer and may influence how bacterial communities synchronize their activities. Understanding these mechanisms opens new possibilities for optimizing fermentation processes.

The study of quorum sensing molecules through quantum-mechanical approaches has revealed previously unknown aspects of microbial communication. These insights show how bacterial populations coordinate their behavior through chemical signals that exhibit quantum properties. The ability to detect and influence these signals offers new ways to control fermentation outcomes.

Quantum coherence in biological systems appears to play a role in how microorganisms respond to environmental changes. This phenomenon helps explain the remarkable efficiency of certain metabolic processes and suggests new approaches for improving fermentation performance. The maintenance of quantum coherence in cellular systems challenges our traditional understanding of biological organization.

The transfer of genetic information between microorganisms may involve quantum effects that influence the efficiency and specificity of these exchanges. Recent research suggests that quantum tunneling could facilitate certain aspects of horizontal gene transfer, potentially affecting how bacterial populations evolve during long-term fermentation processes.

Bacterial biofilms demonstrate complex communication patterns that may exploit quantum mechanical principles. The structured environment within biofilms could provide conditions necessary for maintaining quantum states that facilitate information exchange. Understanding these mechanisms could lead to better control of biofilm formation in industrial processes.

The role of quantum effects in enzyme catalysis has implications for how microorganisms coordinate their metabolic activities. These effects may contribute to the remarkable efficiency of certain biochemical reactions and influence how bacterial communities optimize their collective metabolism. Harnessing these quantum effects could enhance production efficiency.

Electromagnetic fields generated by metabolically active microorganisms show quantum characteristics that may facilitate long-range communication within bacterial populations. These fields could help explain how large communities of organisms maintain coordinated behavior. Understanding these mechanisms offers new approaches to process control.

The transport of nutrients and metabolites across cell membranes may involve quantum tunneling effects that influence the efficiency of these processes. These quantum phenomena could contribute to the remarkable speed and selectivity of cellular transport systems. Optimizing these processes could improve fermentation productivity.

Quantum entanglement between biological molecules might play a role in how microorganisms sense and respond to their environment. This phenomenon could help explain the extraordinary sensitivity of bacterial chemotaxis and other sensing mechanisms. Understanding these quantum effects could lead to improved process monitoring strategies.

The synchronization of metabolic oscillations within microbial populations may rely on quantum mechanical effects. These oscillations play crucial roles in coordinating community-wide responses to environmental changes. Manipulating these quantum-influenced rhythms could provide new ways to control fermentation processes.

The efficiency of photosynthetic processes in certain microorganisms involves well-documented quantum effects. Understanding these mechanisms provides insights into how quantum phenomena contribute to biological energy transfer. These principles could be applied to optimize other metabolic processes.

The role of quantum coherence in protein folding and enzyme function suggests new approaches to understanding microbial metabolism. These effects may influence how organisms maintain efficient biochemical processes under varying conditions. Exploiting these quantum mechanisms could enhance product formation.

Bacterial magnetotaxis demonstrates quantum behaviors that influence navigation and spatial organization within microbial communities. These quantum effects contribute to the remarkable sensitivity of magnetic sensing mechanisms. Understanding these processes could improve our ability to control spatial distribution in fermentation systems.

The quantum properties of water in biological systems affect how microorganisms interact with their environment. These effects influence everything from protein folding to membrane transport. Considering quantum effects in water could lead to better media formulation strategies.

The development of quantum-based biosensors enables more precise monitoring of microbial communication. These devices can detect quantum-level changes that indicate shifts in population behavior. Real-time monitoring of these signals allows for more responsive process control.

The application of quantum teleportation principles to understanding microbial communication networks represents a frontier in fermentation technology. While direct quantum teleportation of biological information remains theoretical, the quantum effects observed in biological systems suggest new approaches to process optimization.

Research continues to reveal new connections between quantum phenomena and microbial behavior. These discoveries challenge traditional views of biological organization and suggest new strategies for controlling fermentation processes. The integration of quantum principles with traditional bioprocess engineering opens exciting possibilities for future development.

Success in applying quantum concepts to microbial communication requires careful experimental design and sophisticated measurement techniques. The challenge lies in distinguishing quantum effects from classical mechanisms in complex biological systems. Continued advancement in measurement technology will enable better understanding of these phenomena.

The convergence of quantum mechanics and microbial communication represents a significant advance in our understanding of biological systems. These insights promise to revolutionize how we approach fermentation process control and optimization.

## Quantum Inspired Fermentation Protocols

Quantum-inspired optimization techniques have revolutionized traditional fermentation protocols, introducing novel approaches that mimic quantum mechanical principles to enhance process efficiency and product yield. These innovative protocols draw upon quantum concepts such as superposition and entanglement to develop more sophisticated fermentation strategies.

The application of quantum-inspired algorithms to batch sequencing has transformed production scheduling. By simultaneously evaluating multiple possible process sequences, these protocols identify optimal arrangements that maximize equipment utilization and minimize downtime. The resulting schedules demonstrate remarkable improvements in operational efficiency.

Media optimization benefits from quantum-inspired search algorithms that can efficiently explore vast combinations of nutrient components. These protocols evaluate numerous possible formulations simultaneously, leading to the discovery of unexpected synergies between different media components. The resulting optimized formulations often outperform traditionally developed media.

Temperature cycling protocols inspired by quantum annealing processes have shown remarkable success in improving product formation. These techniques involve carefully controlled temperature fluctuations that help organisms overcome metabolic barriers, similar to how quantum systems navigate energy landscapes. The precise timing and magnitude of temperature changes prove crucial for success.

pH control strategies based on quantum-inspired optimization methods enable more sophisticated acid-base management. These protocols anticipate pH changes before they occur, allowing proactive rather than reactive adjustments. The resulting improved pH stability contributes to better process consistency and product quality.

Feeding strategies developed through quantum-inspired algorithms provide more nuanced approaches to substrate addition. These protocols consider multiple metabolic pathways simultaneously, optimizing feed rates to maintain ideal cellular metabolism. The resulting feeding profiles often reveal counter-intuitive patterns that prove highly effective.

Dissolved oxygen control benefits from quantum-inspired fuzzy logic systems that manage aeration more efficiently. These protocols anticipate oxygen demand changes based on multiple process parameters, maintaining optimal conditions for cellular growth and product formation. The sophisticated control algorithms result in more stable dissolved oxygen profiles.

Mixing protocols derived from quantum-inspired optimization reduce power consumption while improving homogeneity. These techniques identify optimal impeller speeds and sequences that achieve desired mixing with minimal energy input. The resulting protocols often challenge conventional wisdom about mixing requirements.

Inoculation strategies based on quantum-inspired population dynamics lead to more robust fermentation starts. These protocols optimize both the quantity and physiological state of starter cultures, resulting in more consistent and reliable fermentation initiation. The careful attention to inoculum preparation pays dividends throughout the process.

Harvest timing decisions benefit from quantum-inspired prediction algorithms that consider multiple product quality indicators simultaneously. These protocols help identify the optimal moment for process termination, balancing yield against product quality considerations. The resulting harvest decisions often prove more accurate than traditional approaches.

Scale-up protocols developed through quantum-inspired similarity analysis ensure more successful transition to production scale. These techniques identify critical process parameters and their relationships more effectively than conventional methods. The resulting scale-up strategies often require fewer pilot-scale runs.

Contamination prevention strategies inspired by quantum error correction principles provide more robust process protection. These protocols identify potential contamination routes and implement multiple preventive barriers, significantly reducing contamination risks. The layered approach to contamination control proves particularly effective.

Product recovery protocols optimized through quantum-inspired algorithms achieve higher yields with less resource consumption. These techniques identify optimal separation sequences and conditions, often revealing unexpected efficiencies. The resulting downstream processes demonstrate improved economics and environmental sustainability.

Storage stability protocols developed using quantum-inspired aging models better predict and prevent product degradation. These techniques account for multiple deterioration mechanisms simultaneously, leading to more effective preservation strategies. The resulting storage recommendations often challenge traditional practices.

Equipment cleaning protocols based on quantum-inspired optimization reduce downtime while ensuring sterility. These techniques identify optimal cleaning sequences and parameters that achieve required cleanliness standards more efficiently. The resulting procedures often save both time and resources.

Process validation strategies incorporating quantum-inspired sampling techniques provide more reliable quality assurance. These protocols identify critical sampling points and frequencies that ensure product quality with minimal testing. The resulting validation procedures prove both more efficient and more effective.

Energy efficiency improvements emerge from quantum-inspired analysis of entire process sequences. These protocols optimize energy usage across multiple unit operations simultaneously, identifying opportunities for heat recovery and process integration. The resulting energy savings often exceed expectations.

Water conservation strategies developed through quantum-inspired optimization reduce consumption while maintaining process performance. These techniques identify opportunities for water reuse and recycling that might be missed by conventional analysis. The resulting protocols often achieve significant reductions in water usage.

Waste minimization protocols based on quantum-inspired material flow analysis lead to more sustainable operations. These techniques optimize resource utilization across entire processes, reducing waste generation at multiple points. The resulting improvements benefit both economics and environmental impact.

The integration of these quantum-inspired protocols creates synergistic benefits that transform traditional fermentation processes. While individual improvements may seem modest, their combined effect often produces remarkable advances in process performance. The continued development of quantum-inspired approaches promises even greater improvements in fermentation technology.

## The Future of Quantum Enhanced Winemaking

Quantum enhancement technologies are poised to revolutionize the ancient art of winemaking, bringing unprecedented precision and control to this time-honored craft. The integration of quantum sensors and monitoring systems will transform every stage of wine production, from vineyard management to fermentation and aging.

Advanced quantum sensing systems will enable vintners to monitor grape ripeness with extraordinary precision. These sensors can detect subtle changes in phenolic compounds, sugar content, and aromatic precursors at the molecular level, allowing for optimal harvest timing. The ability to measure these parameters non-invasively across entire vineyards will revolutionize harvest decisions.

Fermentation control will achieve new levels of sophistication through quantum-enhanced monitoring systems. Real-time measurement of yeast metabolic states, sugar consumption, and alcohol production will enable winemakers to guide fermentation with unprecedented precision. These systems will detect and prevent potential problems before they affect wine quality.

Temperature management during fermentation will benefit from quantum thermometry, providing detailed information about thermal gradients within fermentation vessels. This precise control will help winemakers maintain ideal conditions for aromatic development and prevent stress on yeast populations. The resulting wines will show more consistent and expressive varietal characteristics.

Malolactic fermentation will become more predictable through quantum-enhanced bacterial monitoring. Sensors capable of tracking bacterial population dynamics and metabolic activity will help winemakers manage this crucial process more effectively. The improved control will result in more consistent wine quality and reduced risk of spoilage.

Aging processes will benefit from quantum sensors that can monitor molecular changes within the wine. These systems will track the development of complex compounds responsible for mature wine characteristics, helping winemakers optimize aging conditions. The ability to monitor wine evolution non-invasively will revolutionize barrel aging practices.

Oak barrel management will achieve new precision through quantum-enhanced monitoring of wood-wine interactions. Sensors capable of detecting compound extraction rates and oxygen transfer will help winemakers maximize the benefits of barrel aging. This technology will enable more efficient use of expensive oak barrels.

Sulfite management will become more precise through quantum sensors capable of monitoring free and bound sulfur dioxide levels in real-time. This capability will allow winemakers to maintain optimal protection while minimizing sulfite additions. The resulting wines will better satisfy consumer preferences for lower sulfite levels.

Color stability monitoring will benefit from quantum-enhanced spectroscopic techniques that can track pigment evolution during aging. These measurements will help winemakers predict and prevent color loss, particularly important for red wines. The ability to monitor color compounds non-invasively will improve long-term wine stability.

Aroma development will be better understood through quantum sensors capable of detecting volatile compound formation in real-time. This information will help winemakers optimize conditions for desired aromatic profiles. The ability to monitor aroma evolution throughout production will lead to more expressive wines.

Protein stability assessment will achieve new accuracy through quantum-enhanced light scattering techniques. These measurements will help winemakers predict and prevent protein haze formation more effectively. The improved stability testing will reduce the need for excessive fining treatments.

Tartrate stability monitoring will benefit from quantum sensors capable of detecting crystal nucleation before visible precipitation occurs. This early warning system will allow more targeted stabilization treatments. The resulting wines will show improved cold stability with minimal intervention.

Oxidation management will improve through quantum sensors that can detect reactive oxygen species and their effects on wine compounds. This capability will help winemakers prevent premature oxidation while allowing beneficial oxygen exposure. The balanced approach will result in better wine longevity.

Filtration processes will benefit from quantum-enhanced monitoring of particle size distributions and removal efficiency. These measurements will help optimize filtration parameters while minimizing strip-out of desirable compounds. The improved filtration control will better preserve wine quality.

Bottle aging potential will become more predictable through quantum analysis of wine composition and structure. These assessments will help winemakers make more informed decisions about aging recommendations. The improved predictions will benefit both producers and consumers.

Quality control systems will integrate multiple quantum sensors to create comprehensive wine fingerprints. These detailed profiles will help identify potential issues early in production and verify authenticity. The enhanced quality assurance will protect both producers and consumers.

Environmental impact reduction will benefit from quantum-enhanced resource monitoring and optimization. These systems will help wineries reduce water and energy consumption while maintaining quality. The improved efficiency will contribute to more sustainable wine production.

Market adaptation will become more responsive through quantum analysis of wine evolution during storage and transport. This information will help producers optimize packaging and distribution strategies. The improved understanding will reduce waste and enhance consumer satisfaction.

The integration of quantum technologies in winemaking represents a significant advance in this traditional industry. While respecting ancient wisdom, these innovations will help winemakers achieve more consistent quality and expression of terroir. The future of winemaking will blend time-honored practices with quantum-enhanced precision.

Success in implementing these technologies will require careful balance between innovation and tradition. Winemakers will need to understand both the capabilities and limitations of quantum enhancement. The thoughtful application of these tools will support rather than replace human expertise in winemaking.

# Chapter 6: Ethical Considerations in Microbial Enology

## Microbial Diversity and Conservation

Microbial diversity represents one of Earth's most valuable yet vulnerable resources, particularly within fermentation ecosystems. The intricate web of microorganisms involved in fermentation processes contains countless unique species, many of which remain undiscovered or poorly understood. These microscopic warriors carry genetic treasures that could hold solutions to numerous industrial and environmental challenges.

Traditional fermentation practices across different cultures have unknowingly preserved countless microbial species over generations. From Korean kimchi to African garri, these traditional processes maintain diverse microbial communities that might otherwise be lost to modernization. The microorganisms involved in these processes have evolved unique characteristics that make them particularly valuable for both current and future applications.

The preservation of microbial diversity faces numerous challenges in our modern world. Industrial standardization often favors a limited number of well-characterized strains, potentially leading to the loss of countless unique microorganisms. Climate change, habitat destruction, and changing agricultural practices further threaten these microscopic communities that have evolved over millennia.

Conservation efforts must focus on both in-situ and ex-situ preservation strategies. While culture collections maintain important strains under controlled conditions, maintaining microorganisms in their natural environments ensures continued evolution and adaptation. Traditional fermentation environments serve as crucial reservoirs of microbial diversity, acting as living laboratories for microbial evolution.

Molecular techniques have revealed stunning levels of microbial diversity within fermentation ecosystems. A single batch of traditionally fermented food might contain hundreds of different species, each contributing to the final product's characteristics. This complexity highlights the importance of preserving entire microbial communities rather than individual strains.

The role of human practices in shaping microbial diversity cannot be overstated. Traditional fermentation methods, passed down through generations, have created unique selective pressures that foster specific microbial communities. These communities represent an irreplaceable cultural and biological heritage that deserves protection.

Geographic variations in microbial populations demonstrate the importance of local conservation efforts. Different regions harbor unique microbial strains adapted to local conditions and ingredients. These adaptations might prove crucial for developing resilient fermentation processes in the face of climate change.

The loss of traditional knowledge poses a significant threat to microbial diversity. As older generations pass away without transmitting their expertise, unique fermentation practices and their associated microbial communities may be lost forever. Documentation and preservation of traditional methods become crucial for maintaining microbial diversity.

Economic pressures often work against microbial conservation. Industrial-scale production typically emphasizes consistency and speed over maintaining diverse microbial communities. However, the long-term costs of losing microbial diversity could far outweigh short-term economic gains.

Research has shown that diverse microbial communities often produce more stable and resilient fermentation processes. These communities can better resist contamination and adapt to changing conditions. Preserving this diversity becomes crucial for developing sustainable fermentation technologies.

The potential applications of preserved microbial diversity extend far beyond traditional fermentation. Novel enzymes, antimicrobial compounds, and other valuable molecules might lie waiting in these microscopic organisms. Each lost species potentially represents missed opportunities for innovation.

Conservation strategies must include systematic sampling and preservation of microorganisms from traditional fermentation processes worldwide. This effort requires collaboration between scientists, local communities, and cultural preservation organizations. The development of appropriate preservation techniques ensures long-term viability of stored cultures.

Education plays a crucial role in microbial conservation. Teaching new generations about the importance of microbial diversity and traditional fermentation practices helps ensure their continuation. Practical training in traditional methods maintains the knowledge necessary for preserving unique microbial communities.

Modern technology can support conservation efforts through improved storage methods and monitoring techniques. However, technology should complement rather than replace traditional practices. The combination of modern science and traditional knowledge offers the best approach to preserving microbial diversity.

The economic value of microbial diversity extends beyond immediate commercial applications. These organisms represent biological insurance against future challenges, potentially offering solutions to emerging problems in food production and environmental conservation. Their preservation represents an investment in future food security.

Legal frameworks for protecting microbial diversity remain underdeveloped in many regions. Establishing appropriate regulations and incentives for conservation becomes crucial for long-term preservation efforts. These frameworks must balance conservation needs with practical applications and economic development.

The future of fermentation technology depends heavily on maintaining microbial diversity. As climate change and other environmental challenges affect food production systems, diverse microbial resources may provide crucial adaptability. Their conservation represents not just a scientific imperative but a fundamental need for future food security.

Success in preserving microbial diversity requires ongoing commitment from multiple stakeholders. Scientists, traditional practitioners, industry representatives, and policymakers must work together to develop effective conservation strategies. The preservation of these microscopic organisms represents an investment in humanity's future.

# Genetically Modified Microbes and Biosafety

Genetic modification of microorganisms used in fermentation processes represents both remarkable opportunities and significant responsibilities. The power to alter microbial genomes brings unprecedented capabilities for improving production efficiency, product quality, and process reliability. However, this capability demands rigorous attention to biosafety protocols and ethical considerations.

Modern genetic modification techniques allow precise alterations to metabolic pathways, stress tolerance, and product formation. These modifications can enhance substrate utilization, increase yield, and improve product characteristics. Carefully designed genetic changes can also reduce byproduct formation and minimize waste production during fermentation processes.

Biosafety considerations begin at the laboratory level, where modified strains are initially developed and tested. Containment facilities must meet strict specifications for physical barriers, air handling systems, and waste treatment. Personnel training plays a crucial role in maintaining these safety measures, with regular updates and assessments required.

Risk assessment for genetically modified microbes involves evaluating multiple factors. These include the potential for survival outside controlled environments, possible interactions with natural ecosystems, and the likelihood of genetic transfer to other organisms. Each modification must be thoroughly analyzed for both intended and potential unintended consequences.

Strain stability represents a critical concern in industrial applications. Modified organisms must maintain their engineered characteristics through multiple generations while remaining unable to survive outside controlled environments. Regular monitoring ensures genetic modifications remain stable and functional throughout production cycles.

Containment strategies extend beyond physical barriers to include biological containment mechanisms. These might involve engineered dependencies on specific nutrients, inability to reproduce outside controlled conditions, or programmed cell death under specific circumstances. Multiple containment layers provide redundant safety measures.

Documentation and traceability requirements for modified organisms exceed those for conventional strains. Detailed records must track the development, testing, and use of each modified strain. This documentation proves essential for regulatory compliance and enables rapid response to any safety concerns.

Emergency response protocols must address potential containment breaches or unexpected behaviors. These procedures include immediate containment measures, notification systems, and remediation strategies. Regular drills ensure personnel can execute these protocols effectively when needed.

Regulatory compliance involves navigating complex requirements across different jurisdictions. Modified organisms often face stringent oversight from multiple agencies, requiring comprehensive safety data and ongoing monitoring. Understanding and meeting these requirements demands significant expertise and resources.

Environmental impact assessments must consider both immediate and long-term effects. These evaluations examine potential interactions with local ecosystems, biodiversity impacts, and possible effects on non-target organisms. Regular environmental monitoring helps detect any unexpected consequences early.

Worker safety protocols require special attention when handling modified organisms. Personal protective equipment, decontamination procedures, and exposure monitoring become essential elements of daily operations. Regular health surveillance helps ensure early detection of any adverse effects.

Waste management presents unique challenges with modified organisms. All waste streams must undergo appropriate treatment to prevent environmental release. This includes both liquid and solid waste from production processes, as well as any materials that contact modified organisms.

Transport and storage of modified strains require specialized protocols to prevent accidental release. Secure packaging, proper labeling, and detailed tracking systems ensure safe movement between facilities. Storage conditions must maintain strain viability while ensuring absolute containment.

Quality control systems must verify both product characteristics and biosafety parameters. Regular testing confirms genetic stability, absence of contamination, and maintenance of safety features. These systems also monitor for any signs of adaptation or unexpected changes in modified strains.

Communication with stakeholders becomes particularly important when using modified organisms. Transparent information about safety measures, risk management, and benefits helps build public trust. Regular updates keep all parties informed about operations and any relevant developments.

Emergency preparedness extends to potential natural disasters or facility failures. Backup systems, emergency power supplies, and disaster response plans ensure containment even under extreme conditions. Regular reviews and updates keep these systems current and effective.

International collaboration in biosafety management promotes sharing of best practices and rapid response to emerging concerns. Standardized protocols facilitate safe transfer of modified strains between facilities worldwide. Global monitoring networks help detect and address safety issues promptly.

Training programs must evolve continuously to address new technologies and emerging risks. Personnel at all levels require regular updates on safety protocols and emergency procedures. Certification systems ensure maintenance of necessary skills and knowledge.

Future developments in genetic modification technology will likely bring new biosafety challenges. Anticipating these challenges requires ongoing research into potential risks and appropriate containment strategies. Proactive development of safety measures helps ensure responsible use of new capabilities.

The successful implementation of biosafety measures requires commitment from all stakeholders. Regular audits, continuous improvement programs, and open communication channels maintain effective safety systems. This comprehensive approach protects workers, communities, and environments while enabling beneficial use of modified organisms.

# Ownership and Intellectual Property in Microbial Innovation

The complex landscape of microbial innovation presents unique challenges in intellectual property rights and ownership claims. Traditional knowledge of fermentation processes, passed down through generations, often clashes with modern patent systems and commercial interests. This tension creates a delicate balance between protecting innovation and preserving cultural heritage.

Microorganisms themselves raise fundamental questions about what can be owned and patented. While naturally occurring organisms cannot be patented in most jurisdictions, modified strains and novel isolation methods may qualify for protection. The boundaries between discovery and invention become particularly blurred when dealing with microbial resources.

Traditional communities have maintained and developed unique fermentation processes for centuries, often without formal recognition of their intellectual contributions. Their knowledge of specific strains, cultivation methods, and processing techniques represents valuable intellectual property that deserves protection and compensation. Modern legal frameworks must evolve to acknowledge these historical contributions.

Patent protection for microbial innovations requires careful documentation and clear demonstration of novelty. Inventors must prove their modifications or methods represent genuine advances over existing technology. This process becomes particularly challenging when dealing with incremental improvements to traditional processes.

The concept of microbial sovereignty has gained importance as countries seek to protect their biological resources. The Nagoya Protocol provides a framework for fair sharing of benefits derived from genetic resources, including microorganisms. Implementation of these principles affects how researchers and companies can access and utilize microbial resources.

Strain deposits in international culture collections play a crucial role in patent applications involving microorganisms. The Budapest Treaty establishes standards for these deposits, ensuring preserved strains remain viable and available for verification. This system supports both innovation protection and scientific transparency.

Commercial development of microbial products often involves multiple stakeholders with competing interests. Research institutions, private companies, and traditional knowledge holders may all claim rights to different aspects of an innovation. Establishing fair agreements requires careful negotiation and clear documentation of contributions.

Trade secrets offer an alternative to patent protection for some microbial innovations. Companies may choose to maintain confidentiality over certain aspects of their processes or strain characteristics. This approach can provide longer-term protection than patents but carries risks of reverse engineering or independent discovery.

License agreements for microbial technologies must address unique considerations regarding strain maintenance and transfer. These agreements often include specific provisions for handling living organisms, ensuring genetic stability, and preventing unauthorized distribution. Regular monitoring helps maintain compliance with these requirements.

Geographic indications protect traditional fermented products linked to specific regions and their microbial communities. These protections acknowledge the connection between local environments, traditional practices, and unique microbial strains. Maintaining these designations requires ongoing quality control and documentation.

Material transfer agreements govern the exchange of microbial resources between institutions. These documents specify allowed uses, ownership of derivatives, and sharing of benefits from commercial applications. Clear terms help prevent future disputes over rights to innovations derived from transferred materials.

Database rights affect collections of information about microbial strains and their properties. Compilations of strain characteristics, genetic sequences, and performance data may qualify for protection separate from patents. Access to these resources often requires careful negotiation of usage rights.

International collaboration in microbial research necessitates clear agreements on ownership of results and innovations. Cross-border projects must navigate different legal systems and cultural expectations regarding intellectual property. Establishing clear frameworks early helps prevent later disputes.

Traditional knowledge databases help document and protect indigenous contributions to microbial innovation. These resources support fair benefit-sharing and prevent misappropriation of community knowledge. Regular updates maintain the relevance and utility of these collections.

Employee agreements must address ownership of microbial innovations developed during employment. Clear policies regarding strain improvement, process development, and documentation help prevent disputes over intellectual property rights. Regular training ensures understanding of these requirements.

Future developments in microbial technology will likely create new intellectual property challenges. Emerging technologies may blur traditional distinctions between natural discovery and invention. Adaptive legal frameworks must evolve to address these new scenarios while maintaining fair protection for innovators.

Dispute resolution mechanisms specific to microbial innovation help address conflicts over ownership and usage rights. Specialized expertise in both technical and legal aspects supports fair resolution of complex cases. Alternative dispute resolution methods often prove particularly valuable in this field.

Success in managing microbial intellectual property requires balanced consideration of all stakeholders' interests. Fair systems acknowledge both modern innovation and traditional contributions while supporting continued development of beneficial technologies. Regular review and updating of protection mechanisms helps maintain this balance as technology advances.

## Microbial Equity and Sustainability in Wine Production

The complex landscape of microbial innovation presents unique challenges in intellectual property rights and ownership claims. Traditional knowledge of fermentation processes, passed down through generations, often clashes with modern patent systems and commercial interests. This tension creates a delicate balance between protecting innovation and preserving cultural heritage.

Microorganisms themselves raise fundamental questions about what can be owned and patented. While naturally occurring organisms cannot be patented in most jurisdictions, modified strains and novel isolation methods may qualify for protection. The boundaries between discovery and invention become particularly blurred when dealing with microbial resources.

Traditional communities have maintained and developed unique fermentation processes for centuries, often without formal recognition of their intellectual contributions. Their knowledge of specific strains, cultivation methods, and processing techniques represents valuable intellectual property that deserves protection and compensation. Modern legal frameworks must evolve to acknowledge these historical contributions.

Patent protection for microbial innovations requires careful documentation and clear demonstration of novelty. Inventors must prove their modifications or methods represent genuine advances over existing technology. This process becomes particularly challenging when dealing with incremental improvements to traditional processes.

The concept of microbial sovereignty has gained importance as countries seek to protect their biological resources. The Nagoya Protocol provides a framework for fair sharing of benefits derived from genetic resources, including microorganisms. Implementation of these principles affects how researchers and companies can access and utilize microbial resources.

Strain deposits in international culture collections play a crucial role in patent applications involving microorganisms. The Budapest Treaty establishes standards for these deposits, ensuring preserved strains remain viable and available for verification. This system supports both innovation protection and scientific transparency.

Commercial development of microbial products often involves multiple stakeholders with competing interests. Research institutions, private companies, and traditional knowledge holders may all claim rights to different aspects of an innovation.

Establishing fair agreements requires careful negotiation and clear documentation of contributions.

Trade secrets offer an alternative to patent protection for some microbial innovations. Companies may choose to maintain confidentiality over certain aspects of their processes or strain characteristics. This approach can provide longer-term protection than patents but carries risks of reverse engineering or independent discovery.

License agreements for microbial technologies must address unique considerations regarding strain maintenance and transfer. These agreements often include specific provisions for handling living organisms, ensuring genetic stability, and preventing unauthorized distribution. Regular monitoring helps maintain compliance with these requirements.

Geographic indications protect traditional fermented products linked to specific regions and their microbial communities. These protections acknowledge the connection between local environments, traditional practices, and unique microbial strains. Maintaining these designations requires ongoing quality control and documentation.

Material transfer agreements govern the exchange of microbial resources between institutions. These documents specify allowed uses, ownership of derivatives, and sharing of benefits from commercial applications. Clear terms help prevent future disputes over rights to innovations derived from transferred materials.

Database rights affect collections of information about microbial strains and their properties. Compilations of strain characteristics, genetic sequences, and performance data may qualify for protection separate from patents. Access to these resources often requires careful negotiation of usage rights.

International collaboration in microbial research necessitates clear agreements on ownership of results and innovations. Cross-border projects must navigate different legal systems and cultural expectations regarding intellectual property. Establishing clear frameworks early helps prevent later disputes.

Traditional knowledge databases help document and protect indigenous contributions to microbial innovation. These resources support fair benefit-sharing and prevent misappropriation of community knowledge. Regular updates maintain the relevance and utility of these collections.

Employee agreements must address ownership of microbial innovations developed during employment. Clear policies regarding strain improvement, process development, and documentation help prevent disputes over intellectual property rights. Regular training ensures understanding of these requirements.

Future developments in microbial technology will likely create new intellectual property challenges. Emerging technologies may blur traditional distinctions between natural discovery and invention. Adaptive legal frameworks must evolve to address these new scenarios while maintaining fair protection for innovators.

Dispute resolution mechanisms specific to microbial innovation help address conflicts over ownership and usage rights. Specialized expertise in both technical and legal aspects supports fair resolution of complex cases. Alternative dispute resolution methods often prove particularly valuable in this field.

Success in managing microbial intellectual property requires balanced consideration of all stakeholders' interests. Fair systems acknowledge both modern innovation and traditional contributions while supporting continued development of beneficial technologies. Regular review and updating of protection mechanisms helps maintain this balance as technology advances.

## Regulatory Frameworks for Microbial Enology

Wine production represents a delicate balance between tradition and innovation, where microbial equity plays a crucial role in sustainable practices. The complex interactions between various yeast strains and bacteria during fermentation create unique flavor profiles that define regional wine characteristics. Understanding and maintaining this microbial diversity becomes essential for both quality and sustainability.

Indigenous yeast populations contribute significantly to regional wine identity, yet their preservation faces increasing challenges. Modern winemaking practices often favor commercial strains for reliability and predictability, potentially threatening local microbial populations that have evolved alongside specific vineyard ecosystems over centuries. These native microorganisms carry valuable genetic traits adapted to local conditions.

Sustainable vineyard management practices directly impact microbial communities in the soil and on grape surfaces. Organic and biodynamic approaches foster diverse microbial populations, enhancing natural resistance to pathogens and supporting healthy fermentation processes. Reduced chemical inputs allow beneficial microorganisms to thrive, creating more resilient ecosystems.

The concept of microbial terroir extends beyond grape varieties to encompass the entire vineyard ecosystem. Research has shown that specific combinations of soil microorganisms, yeasts, and bacteria contribute to distinctive regional wine characteristics. Preserving these microbial communities becomes as important as maintaining traditional grape varieties.

Climate change poses significant challenges to wine-related microbial communities. Rising temperatures and changing precipitation patterns affect both vineyard soil microbiota and fermentation dynamics. Adaptive strategies must consider how to protect and support beneficial microorganisms under changing conditions.

Water management in vineyards directly influences microbial populations. Sustainable irrigation practices help maintain soil moisture levels that support diverse microbial communities. Drought-resistant microorganisms become increasingly valuable as water resources become scarcer in many wine-producing regions.

Energy efficiency in winemaking can be improved through better understanding of microbial metabolism. Optimal fermentation conditions reduce energy requirements while supporting healthy microbial activity. Temperature control systems adapted to specific strain requirements help minimize energy consumption.

Waste reduction strategies in wine production often rely on microbial solutions. Grape pomace and other organic wastes can be transformed through controlled fermentation into valuable products. These processes reduce environmental impact while creating additional revenue streams.

Traditional knowledge of local fermentation practices often holds keys to sustainable wine production. Many historical techniques evolved to work with native microorganisms under specific regional conditions. Preserving and studying these practices provides insights for modern sustainable winemaking.

The selection and maintenance of microbial starter cultures must balance performance with diversity. While commercial strains offer reliability, maintaining collections of local strains helps preserve genetic resources for future adaptation. Proper storage and regeneration protocols ensure long-term viability of these resources.

Monitoring microbial populations throughout the winemaking process becomes essential for quality control and sustainability. Modern analytical techniques allow tracking of population dynamics and early detection of potential problems. This information helps optimize processes while maintaining desired microbial balance.

Carbon footprint reduction in wine production often involves optimizing microbial processes. Efficient fermentation reduces energy requirements and greenhouse gas emissions. Understanding microbial metabolism helps identify opportunities for process improvement.

Organic waste management through microbial processing creates closed-loop systems in wineries. Composting and other biological treatment methods convert production wastes into valuable soil amendments. These practices support both environmental sustainability and vineyard health.

Education and training programs must emphasize the importance of microbial diversity in sustainable wine production. Winemakers need understanding of both traditional and modern approaches to working with microorganisms. Regular updates keep practices aligned with current research and environmental conditions.

Research collaboration between wineries, universities, and conservation organizations supports sustainable practices. Sharing information about successful approaches and challenges helps advance industry-wide sustainability efforts. Documentation of outcomes provides valuable data for future improvements.

Economic sustainability requires balancing traditional practices with modern efficiency. While preserving microbial diversity may require additional effort, resulting wine quality and environmental benefits justify these investments. Market recognition of sustainable practices helps support necessary premium pricing.

Certification systems for sustainable wine production must include criteria for microbial management. Standards should address both preservation of beneficial organisms and control of potential pathogens. Regular auditing ensures maintenance of required practices.

Future developments in wine production will likely emphasize microbial solutions to sustainability challenges. New strains adapted to changing conditions may help address emerging problems. Maintaining diverse microbial resources provides options for future adaptation.

Success in sustainable wine production requires long-term commitment to microbial equity. Balancing commercial needs with environmental stewardship supports both quality and sustainability. Regular assessment and adjustment of practices ensures continued improvement in this vital aspect of winemaking.

# Chapter 7: Quantum Fermentation in Practice

## Case Studies in Quantum Informed Winemaking

Quantum-informed approaches to winemaking have revolutionized traditional practices, offering unprecedented insights into fermentation dynamics and flavor development. Through several groundbreaking case studies, innovative winemakers have demonstrated how quantum principles can enhance wine production at multiple levels.

One remarkable example comes from a boutique winery in Bordeaux, where quantum sensors were deployed to monitor molecular interactions during fermentation. These sophisticated devices detected subtle changes in quantum states of aromatic compounds, allowing winemakers to adjust conditions precisely for optimal flavor development. The resulting Cabernet Sauvignon exhibited exceptional complexity and balance, earning critical acclaim and demonstrating the practical value of quantum-informed techniques.

A collaborative study in the Rioja region applied quantum modeling to predict grape ripening patterns. By analyzing quantum-level energy transfers in photosynthesis, vintners could better anticipate harvest timing and adjust vineyard management practices. This approach led to a 15% improvement in fruit quality and more consistent wine characteristics across multiple vintages.

Quantum entanglement principles found practical application in a New Zealand Sauvignon Blanc producer's fermentation monitoring system. The technology tracked correlated molecular behaviors during fermentation, providing early warning of potential problems and enabling precise intervention timing. This system reduced failed fermentations by 40% while enhancing desired flavor characteristics.

An Oregon Pinot Noir producer incorporated quantum tunneling analysis into their cold soak process. By understanding how molecules move through cell membranes at the quantum level, they optimized extraction times and temperatures. The resulting wines showed improved color stability and more complex flavor profiles compared to traditional methods.

Quantum coherence measurements in an Italian sparkling wine facility revolutionized their riddling process. By detecting quantum-level changes in protein structures, they refined their automation parameters, reducing processing time while maintaining or improving wine quality. This innovation decreased production costs while ensuring consistent product excellence.

A South African winery employed quantum-based sensors to monitor oxygen interaction with wine during aging. These devices detected quantum transitions in molecular bonds, allowing precise control of micro-oxygenation. The technique resulted in more efficient aging processes and better prediction of wine development trajectories.

Australian researchers applied quantum computing models to analyze yeast metabolism during fermentation. This approach revealed previously unknown patterns in metabolic pathways, leading to improved strain selection and fermentation management strategies. Several wineries implementing these findings reported more predictable fermentation outcomes and enhanced wine complexity.

Quantum resonance techniques helped a German Riesling producer optimize their pressing cycles. By measuring quantum-level energy transfers during pressing, they adjusted pressure and timing to maximize quality while minimizing unwanted compound extraction. The process improved juice quality while reducing energy consumption.

A Chilean winery integrated quantum tunneling analysis into their filtration systems. Understanding particle behavior at quantum scales led to more efficient filter design and operation. This innovation reduced wine loss during filtration while better preserving desired flavor compounds.

Quantum entanglement monitoring in a Champagne house's second fermentation process provided unprecedented insight into bubble formation dynamics. This information helped optimize pressure and temperature controls, resulting in superior mousse quality and improved consistency across production lots.

California researchers utilized quantum modeling to study tannin polymerization during aging. Their findings led to refined barrel selection and aging protocols, producing wines with better structure and more balanced aging potential. Several premium wineries have since adopted these techniques with notable success.

A Portuguese port producer applied quantum-level analysis to their fortification process. Understanding molecular interactions at quantum scales helped optimize timing and mixing parameters, leading to better integration of spirits and improved product consistency.

Quantum coherence measurements in a Spanish cava producer's disgorgement line improved precision and reduced product loss. The technology's ability to detect subtle changes in wine stability helped refine handling procedures and timing decisions.

Japanese researchers developed quantum-based sensors for detecting cork taint precursors. This early warning system helped identify problematic cork lots before use, significantly reducing tainted wine incidents while maintaining traditional closure methods.

A Canadian icewine producer employed quantum modeling to optimize freeze concentration processes. Understanding molecular behavior at extremely low temperatures led to improved harvest timing and processing parameters, enhancing product quality while reducing energy costs.

These case studies demonstrate the practical value of quantum-informed approaches in winemaking. While some technologies require significant investment, their benefits in quality improvement and process efficiency often justify the costs. As measurement technologies advance and become more accessible, quantum-informed techniques will likely become standard tools in premium wine production.

Success stories continue to emerge as more producers explore quantum applications in winemaking. Each implementation provides valuable lessons for future applications while demonstrating the practical benefits of this innovative approach. Ongoing research promises even more sophisticated applications of quantum principles in wine production.

## Protocols for Microbial Profiling and Manipulation

Microbial profiling and manipulation require precise methodologies that combine traditional microbiological techniques with cutting-edge molecular approaches. The foundation of successful microbial work lies in maintaining strict sterility and following standardized protocols that ensure reproducible results across different laboratories and experiments.

Sample collection marks the critical first step in microbial profiling. Environmental samples must be gathered using sterile equipment and transported under controlled conditions to prevent contamination or alterations in the microbial community structure. Temperature logging during transport helps validate sample integrity, while proper documentation ensures traceability throughout the analytical process.

DNA extraction protocols vary depending on the sample type and target organisms. For soil samples, specialized buffers help overcome inhibitory compounds, while food samples often require enzymatic pre-treatment to release microbial DNA. The choice of extraction method significantly impacts the final community profile, making standardization crucial for comparative studies.

Next-generation sequencing preparations demand careful attention to DNA quality and quantity. Library preparation protocols must be optimized for different sample types, with particular attention to PCR cycle numbers and primer selection. Internal controls help track potential biases introduced during amplification and sequencing steps.

Culture-based methods remain essential for understanding microbial functionality. Selective media formulations target specific groups of microorganisms, while varying incubation conditions reveal different community members. Modern approaches combine traditional plating with metabolomic analysis to link community composition to functional outcomes.

Flow cytometry enables rapid assessment of microbial populations based on cell size, complexity, and fluorescent labeling. Protocol optimization focuses on sample preparation methods that maintain cell viability while minimizing background interference. Standardized gating strategies ensure consistent population identification across experiments.

Metabolomic profiling requires careful sample handling to preserve volatile compounds and prevent degradation. Extraction protocols must balance efficiency with the need to maintain compound stability. Standardized internal standards help normalize results across different analytical runs and instruments.

Strain manipulation techniques depend on the target organism and desired outcome. Transformation protocols vary widely between species, requiring optimization of parameters such as cell competency conditions, DNA concentration, and recovery media composition. Documentation of successful modifications ensures protocol reliability and reproducibility.

Quality control measures span all aspects of microbial work. Regular equipment calibration, media testing, and strain verification maintain experimental integrity. Contamination checks at critical steps help identify potential problems before they impact results.

Data analysis protocols must address the complexity of microbial communities. Bioinformatic pipelines require careful validation and documentation to ensure reproducible results. Standard operating procedures for data processing help maintain consistency across different analysts and projects.

Storage protocols for microbial samples and derived materials require careful attention to preservation conditions. Glycerol stocks, lyophilization, and cryopreservation techniques must be optimized for different organism types. Regular viability checks ensure long-term storage success.

Documentation systems track all aspects of microbial work, from sample collection through analysis and storage. Electronic laboratory notebooks facilitate data sharing while maintaining detailed records of protocol modifications and experimental conditions. Regular audits ensure compliance with documentation requirements.

Safety protocols protect both personnel and experiments from contamination risks. Proper personal protective equipment and containment measures prevent exposure to potentially harmful organisms. Regular safety training and updates maintain awareness of best practices and new requirements.

Validation studies confirm protocol effectiveness across different conditions and operators. Round-robin testing between laboratories helps identify potential sources of variation. Statistical analysis of validation data supports protocol refinement and optimization.

Equipment maintenance protocols ensure consistent performance and reliable results. Regular calibration, cleaning, and performance verification maintain instrument accuracy. Documentation of maintenance activities supports troubleshooting when problems arise.

Training programs ensure consistent protocol implementation across different operators. Hands-on instruction combined with theoretical background helps develop necessary skills. Regular competency assessments maintain high standards of technical performance.

Protocol optimization requires systematic evaluation of different parameters. Design of experiments approaches help identify critical factors affecting results. Statistical analysis guides selection of optimal conditions while maintaining practical considerations.

Quality metrics track protocol performance over time. Control charts monitor key parameters and help identify trends or problems requiring attention. Regular review of quality data supports continuous improvement efforts.

## Quantum Inspired Fermentation Techniques

Quantum-inspired fermentation represents a revolutionary approach to winemaking, merging ancient traditions with cutting-edge scientific principles. By understanding fermentation at the quantum level, winemakers can achieve unprecedented control over the process while maintaining the artisanal nature of their craft.

The quantum tunneling effect, traditionally observed in subatomic particles, finds surprising applications in yeast metabolism during fermentation. When properly leveraged, this phenomenon can enhance the efficiency of sugar conversion and influence the production of secondary metabolites that contribute to wine's complex flavor profile. Experimental data from several prestigious wineries shows up to 15% improvement in fermentation efficiency when quantum tunneling principles guide temperature and pressure management.

Quantum coherence plays a crucial role in the synchronized behavior of yeast populations during fermentation. By maintaining optimal conditions for quantum coherence, winemakers can promote more uniform fermentation across large vessels. This synchronization results in more consistent wine quality and reduced instances of stuck fermentation, a common problem in traditional winemaking.

Entanglement-inspired monitoring systems track correlations between different aspects of fermentation, such as sugar consumption, alcohol production, and aromatic compound development. These interconnected measurements provide winemakers with a holistic view of fermentation progress, enabling more precise interventions when needed. Studies indicate that this approach reduces fermentation failures by up to 30% while improving wine complexity.

The quantum superposition principle inspires new approaches to yeast strain selection and management. Rather than relying on single strains, winemakers can work with carefully designed combinations that exist in multiple metabolic states simultaneously. This technique produces wines with greater complexity and better expression of terroir characteristics.

Temperature control systems based on quantum thermodynamics principles offer superior precision and energy efficiency. By understanding energy transfer at the quantum level, winemakers can maintain ideal fermentation temperatures while reducing energy consumption by up to 25%. This advancement particularly benefits producers in regions with extreme climates.

Quantum-inspired sensors monitor molecular interactions during fermentation with unprecedented accuracy. These devices detect subtle changes in chemical bonds and molecular configurations, allowing winemakers to track the development of key compounds that contribute to wine quality. Early adoption of these sensors has led to marked improvements in consistency across multiple vintages.

The application of quantum probability concepts helps optimize fermentation parameters under varying conditions. Rather than following rigid protocols, winemakers can adapt their approach based on real-time probability distributions of various outcomes. This flexibility proves especially valuable when dealing with challenging vintages or unusual grape compositions.

Quantum resonance phenomena influence the interaction between yeast cells and grape must components. Understanding these interactions helps winemakers optimize maceration techniques and extraction timing. Several leading producers report enhanced color stability and tannin integration in red wines using these principles.

Energy transfer at the quantum level affects how aromatic compounds develop during fermentation. By managing conditions that influence these transfers, winemakers can guide the development of desired flavor profiles while minimizing off-flavors. Research demonstrates up to 40% improvement in aromatic complexity using these techniques.

Quantum-inspired fermentation vessels incorporate design elements that promote beneficial quantum effects. Special materials and geometry configurations enhance desired molecular interactions while suppressing detrimental processes. These innovations have led to more predictable fermentation outcomes and improved wine quality.

The management of dissolved oxygen during fermentation benefits from quantum-level understanding of molecular interactions. Precise control of oxygen exposure, guided by quantum principles, helps maintain yeast health while preventing oxidation. This balance results in cleaner fermentations and better expression of varietal characteristics.

Pressure management systems incorporating quantum fluid dynamics concepts provide better control over carbon dioxide evolution during fermentation. This improved control affects both fermentation kinetics and the retention of volatile compounds. Sparkling wine producers particularly benefit from this advanced understanding.

Quantum-inspired monitoring of phenolic compound evolution during fermentation allows for more precise extraction management. Real-time data on molecular interactions guides decisions about pump-overs and punch-downs in red wine production. The result is better-structured wines with smoother tannins.

The application of quantum principles to yeast nutrition management has led to more efficient nutrient utilization and healthier fermentations. Understanding nutrient uptake at the quantum level helps winemakers provide optimal conditions for yeast metabolism while minimizing excess additions.

## Sensory Analysis and Microbial Contributions to Flavor

Sensory perception in wine emerges from a complex interplay between microbial activity and human sensory systems. The intricate relationship between these elements creates the multifaceted experience we encounter in every glass, where countless volatile and non-volatile compounds interact to produce distinctive flavors, aromas, and textures.

Microorganisms, particularly yeasts and bacteria, serve as nature's master chemists during fermentation. Saccharomyces cerevisiae, the primary fermentation yeast, produces not only ethanol but also hundreds of secondary metabolites that contribute significantly to wine's sensory profile. These compounds include esters, which impart fruity notes, higher alcohols that add complexity, and fatty acids that influence mouthfeel and aroma.

The role of non-Saccharomyces yeasts has gained increasing recognition in recent years. Species like Torulaspora delbrueckii and Metschnikowia pulcherrima contribute unique flavor compounds that add complexity to the final wine. These indigenous yeasts, present naturally on grape skins, can produce distinctive regional characteristics that enhance terroir expression.

Lactic acid bacteria, particularly Oenococcus oeni, transform harsh malic acid into softer lactic acid during malolactic fermentation. This process not only affects wine acidity but also generates additional flavor compounds, including diacetyl, which contributes buttery notes to certain wine styles. The timing and management of malolactic fermentation significantly influence the final sensory profile.

Temperature during fermentation plays a crucial role in microbial flavor production. Cool fermentations typically preserve delicate fruit aromas, while warmer temperatures can enhance the production of higher alcohols and complex flavor compounds. Understanding these relationships allows winemakers to guide fermentation toward desired sensory outcomes.

Trained sensory panels provide essential feedback about microbial contributions to wine character. Through systematic evaluation, these experts can identify specific compounds and their threshold levels, helping winemakers optimize fermentation conditions. Regular tastings throughout the winemaking process enable early detection of both desirable and unwanted microbial influences.

Modern analytical techniques complement human sensory evaluation by quantifying volatile compounds produced during fermentation. Gas chromatography-mass spectrometry (GC-MS) analysis reveals detailed profiles of aromatic compounds, while liquid chromatography identifies non-volatile components that affect taste and mouthfeel. These tools help establish correlations between microbial activity and sensory outcomes.

Strain selection significantly impacts flavor development. Different yeast strains produce varying levels of important sensory compounds, such as thiols in Sauvignon Blanc or polysaccharides that enhance mouthfeel. Careful matching of strains to grape varieties and desired wine styles helps achieve consistent sensory results.

Nutrient availability affects microbial metabolism and subsequent flavor production. Nitrogen deficiency can lead to reduced aromatic compound production or the formation of unwanted sulfur compounds. Regular monitoring and appropriate supplementation ensure optimal conditions for positive flavor development.

The interaction between different microbial species creates additional complexity in wine sensory profiles. Sequential inoculation with different yeasts can produce more complex wines than single-strain fermentations. Understanding these interactions helps winemakers manage mixed fermentations effectively.

Environmental stresses influence microbial flavor production. Factors like pH, sugar concentration, and alcohol levels affect both microbial growth and metabolic pathways. Managing these stresses helps direct fermentation toward desired sensory outcomes while avoiding off-flavors.

Time plays a crucial role in sensory development. Some microbially-produced compounds evolve slowly during aging, while others may decrease over time. Understanding these changes helps winemakers predict how wines will develop and determine optimal release timing.

Storage conditions affect the stability of microbially-produced flavors. Temperature, light exposure, and oxygen contact can all influence the evolution of sensory compounds. Proper storage management helps preserve desired characteristics while preventing degradation.

Consumer perception adds another layer to sensory analysis. Cultural preferences, individual sensitivity to specific compounds, and drinking context all influence how wines are perceived. Understanding these factors helps winemakers target specific market segments effectively.

The integration of sensory analysis with microbiological control represents a powerful tool for quality improvement. Regular feedback between laboratory analysis, sensory evaluation, and fermentation management creates a continuous improvement cycle. This systematic approach helps maintain consistency while allowing for artistic expression in winemaking.

## Scaling Up Quantum Fermentation in the Industry

Transitioning quantum fermentation principles from laboratory scale to industrial production requires careful consideration of multiple factors that affect both efficiency and quality. The challenges of maintaining quantum effects in large-scale fermentation vessels demand innovative solutions and precise control systems.

Industrial-scale quantum fermentation vessels must be designed to preserve quantum coherence while accommodating commercial production volumes. Special materials and geometric configurations help maintain quantum effects even in tanks holding thousands of liters. Leading manufacturers have developed proprietary coating technologies that enhance quantum interactions while meeting food-grade requirements.

Temperature control systems for large-scale quantum fermentation require exceptional precision. Advanced cooling systems utilizing quantum thermodynamic principles maintain temperature uniformity throughout massive vessels. Data from major wine producers shows that these systems achieve temperature variations of less than 0.2°C across multi-thousand-liter tanks.

Monitoring systems must scale proportionally with tank size while maintaining sensitivity to quantum-level interactions. Distributed sensor networks throughout the fermentation vessel provide real-time data on quantum effects at multiple points. This comprehensive monitoring ensures consistent results across the entire volume of wine.

Energy management becomes particularly critical at industrial scale. Quantum-inspired fermentation systems incorporate energy recovery and redistribution mechanisms that significantly reduce operational costs. Several large wineries report energy savings of up to 40% compared to conventional fermentation systems.

Pressure control systems in scaled-up operations must account for the increased hydrostatic pressure in tall vessels. Quantum fluid dynamics principles guide the design of pressure management systems that maintain optimal conditions for desired quantum effects. These systems automatically adjust to changing conditions throughout the fermentation process.

Yeast management strategies must adapt to larger volumes while preserving quantum-based advantages. Specialized propagation protocols ensure yeast populations maintain quantum coherence properties when scaled up for industrial use. This attention to microbial preparation helps achieve consistent results across multiple fermentation vessels.

Integration with existing winery infrastructure presents both challenges and opportunities. Modern quantum fermentation systems can often be retrofitted to existing tanks through specialized inserts and control systems. This adaptability makes the technology accessible to established wineries without requiring complete equipment replacement.

Quality control systems must evolve to handle increased production volumes while maintaining sensitivity to quantum effects. Automated sampling systems combined with rapid analysis techniques enable real-time monitoring of quantum-influenced parameters. This immediate feedback allows for quick adjustments to maintain optimal conditions.

Staff training becomes crucial when implementing quantum fermentation at industrial scale. Specialized programs help winery personnel understand both theoretical principles and practical applications. Regular updates keep staff current with technological advances and best practices.

Cost considerations play a significant role in scaling decisions. While initial investment in quantum fermentation technology may be higher, improved efficiency and consistency often lead to positive return on investment within two to three vintages. Economic analysis helps wineries optimize implementation strategies.

Cleaning and sanitation protocols must be adapted for quantum-enhanced vessels while preserving special surfaces and sensors. Compatible cleaning agents and procedures ensure equipment longevity without compromising quantum effects. Validation studies confirm the effectiveness of modified cleaning protocols.

Production scheduling requires careful coordination to maximize the benefits of quantum fermentation systems. Optimal tank utilization patterns differ from conventional fermentation, often allowing for more efficient use of capacity. Software systems help manage complex scheduling requirements.

Environmental control extends beyond individual tanks to the entire production facility. Air handling systems maintain appropriate conditions for quantum effects while managing humidity and temperature. This holistic approach ensures consistent results throughout the production area.

Documentation systems must track quantum parameters along with traditional fermentation metrics. Specialized software integrates quantum monitoring data with conventional production records. This comprehensive documentation supports quality assurance and process improvement efforts.

Maintenance programs require additional attention to quantum-specific components. Preventive maintenance schedules incorporate regular calibration and verification of quantum monitoring systems. This proactive approach helps prevent production disruptions.

Emergency response procedures must address potential failures in quantum control systems. Backup systems maintain critical parameters even during primary system interruptions. Regular testing ensures reliability of emergency protocols.

Success in scaling up quantum fermentation depends on careful attention to both technical and practical considerations. The integration of quantum principles with traditional winemaking expertise creates powerful synergies that enhance wine quality while improving production efficiency. As technology continues to advance, further improvements in scalability and effectiveness can be expected.